Ellis Bros

Descriptive list of plants

Ellis Bros

Descriptive list of plants

ISBN/EAN: 9783741183355

Manufactured in Europe, USA, Canada, Australia, Japa

Cover: Foto ©berggeist007 / pixelio.de

Manufactured and distributed by brebook publishing software
(www.brebook.com)

Ellis Bros

Descriptive list of plants

OUR GREETING.—With this, the twenty-seventh edition of our annual Catalogue, we wish to thank our many friends and customers for their kind words as well as trade favors, and to wish you success in the plans of the year. Our stock for the coming season will be fully up to past standard, and as good as that sent out by any firm in the country.

Again thanking you all for past or coming favors, we remain, respectfully,

KEENE, N. H., FEBRUARY 1, 1899.   ELLIS BROS.

HINTS TO CUSTOMERS WHEN ORDERING.—Write name of each article on separate line; order only what you find described and priced in this Catalogue; keep a copy of order; state how to ship (mail or express); also, do not forget most important of all—to sign plainly, the Town, State, P. O. Box or Street number.

POINTS.—Greenhouses newly built and latest in style for plant growing. Ten houses, with 25,000 feet of glass. All shipping stock grown cool. Packing and shipping by up-to-date methods. Customers in every State and in Canada. Sending out more packages of greenhouse plants by mail than any other firm in New England.

PLANT INDEX.

BUSINESS OPENING ! *To one or more men with fair amount of capital. Address Ellis Bros.*
for particulars, as this will bear investigation.

Descriptive List of Plants.

FERNS.

The popularity and demand for fine Ferns has increased wonderfully in past few years, and they are now indispensable; their diversity and gracefulness of foliage make them peculiarly fitted to combine with other plants and flowers; also extra nice in ferneries and for table decoration. They delight in light sandy loam, mixed with leaf mould, and with good drainage, for while they thrive in a moist soil, stagnant or sour earth is fatal to success. We have endeavored to grow in the greatest variety of foliage the best sorts for the amateur, also, unequalled for the florist. The small plants can go by mail, but larger sizes should be ordered sent by express.

Polystichum Proliferum (or Lace Fern). One of the finest of Ferns for house culture; very strong grower; long, graceful fronds, keeping a long time when cut and used with cut flowers. If grown in a partially shaded place, this will be found one of the most pleasing of house plants. We see that several firms are offering this fine Fern under the new but incorrect name of "Aspidium Proliferum Walastoni." Certainly a grand plant, but no better for its new name. (See cut.)   Price 15 cts.

Pteris Serrulata Voluta. This desirable Fern is one of the most distinct varieties of Pteris Serrulata yet introduced, the volute or curled appearance being very noticeable, not only in the mature plants, but also in the young state. Its elegant appearance being also further enhanced by the tips being prettily crested. A rare Fern and yet offered but by few in this country.   Price 30 cts.

Adiantum Capillus-Veneris Imbricatum. This is one of the grandest additions to the Fern family, and one that is sure to be grown extensively on account of its beauty. The pinnæ are very broad, two or more inches in diameter, deeply fringed and overlapping each other, thus forming a solid yet graceful arching frond of rich green color and good texture. This variety will be found most useful and valuable, not only for collections and decorations, but especially for the filling of jardinieres and fern dishes on account of its dwarf and bushy habit.   It will also serve well for cutting purposes, especially for the boutonnieres and small bouquets. Price 40 cts.

Nephrolepis Cordata Compacta (new Sword Fern). In our estimation this variety is the finest of all the Sword Ferns; it is of free, strong-growing compact habit, attaining when fully grown a height of about two feet. The fronds, which are of a very rich dark green color, of upright growth, with just sufficient arch in them to make them graceful. As a Fern for house culture or for window gardening we do not know of any other variety that would give the same satisfaction and which we could recommend so highly. A beauty, it cannot be too highly praised.   (See cut.)   Price 20 cts., 50 cts. and 75 cts.

Boston Sword Fern (Nephrolepis Exaltata Bostoniensis). A variation originating in the vicinity of Boston, where it has been in popular favor for a number of years under the name of Boston Fern. In the vicinity of Boston no other plant is used so extensively as this graceful Nephrolepis, which differs from the ordinary Fern in having much longer fronds, which frequently attain a length of four feet. These fronds arch and droop over very gracefully, on account of which it is frequently called the Fountain Fern. This drooping habit makes it an excellent fern to grow as a single specimen on a table or pedestal. We pride ourselves in being the first to offer this grand Fern to retail catalogue trade of this country. (See cut.)

Price 20 cts., 50 cts. ; fine specimens, 75 cts.

Pteris Victoria. A most remarkable and beautifully variegated Pteris, and undoubtedly one of the most striking and desirable Ferns of recent introduction. The foliage is neatly divided, the sterile fronds being much wider than the fertile ones, thus giving the plant a most chaste and lovely appearance. The fronds are of rich green color, with beautiful silvery white variegation.   Price 20 cts.

Polystichum Coreaceum. A low growing sort, forming a dense mass of light green foliage; extra good for Fern pans and Fern cases. The fronds of this sort have wonderful keeping qualities when cut, lasting in vases of water when well cared for from six to ten weeks. Easily grown.   Price 15 cts.

Pteris Tremula (or Shaking Fern). One of the finest Ferns for house decoration, growing very rapidly and throwing up large handsome fronds. It makes magnificent specimens. Easily grown and very popular. One of the best Ferns grown. (See cut.)   Price 15 cts. and 35 cts.

Adiantum Cuneatum. The most popular of all the "Maiden Hair" varieties. Fronds delicate and graceful; a distinct, lovely variety. Price 15 cts. and 30 cts.

Cyrtomium Falcatum. Quite a new Fern from Japan, of low ornamental growth, fronds of great substance, thick and leathery, keeping well on plant or when cut. This variety is now very popular and in great demand at Christmas in sections of country where known, it being called the Holly Fern, from resemblance of foliage on mature fronds to that of Holly.   Price 20 cts.

Polypodium Aureum. A grand large growing sort, making a fine specimen in short time. Very noticeable on account of foliage, which is very large and of a distinct bluish color.   Price 20 cts.; larger, 35 cts.

Dicksonia Antartica. Royal and majestic; perhaps the most generally admired of all Tree Ferns. The tall straight column of the trunk has an unusually rich, luxuriant crown of delicately cut but firm-textured fronds. For unique decorative effects the Tree Ferns are unsurpassed. Being rare and not in common use, they attract attention. Young plants do not show the tree form; as they grow older they gradually form a trunk or tree shape.   Price 20 and 35 cts.

Platycerium Grande. Better known as Staghorn Fern, owing to the striking resemblance of the fronds to the horns of a stag.   Price 50 cts. to $3.00 each.

Pteris Argyrea. A very strong growing variety and very useful for all purposes. Quite large foliage, with broad white band through the center of each frond.
                                        Price 20 cts.

PETUNIAS.

Mrs. Morton. This stands at the head of all the white frilled varieties. Growth vigorous, and very dwarf and bushy; flowers white, with very heavy finely cut fringe.

The most light and airy in appearance of any of the whites. Good as a bedder, and an excellent pot plant.   Price 20 cts.

Comet. Large flowers, heavily fringed; color a rich purple, each petal edged white; very fine.   Price 15 cts.

Eureka. Pure white, each petal having a narrow edge of magenta; elegant form; one of the most desirable.   Price 15 cts.

Emdymion. Very delicate color of soft rose, elegantly frilled; a distinct and pleasing sort, desirable in any collection.   Price 15 cts.

Mrs. Sanders. A new variety, of the most perfect form; color a fine pure pink; one of the best of its color.   Price 15 cts.

NOTE.—Above are all fine double fringed sorts.

Petunia. Single, mixed from fine strains of the large flowering sorts.
    Price 10 cts.; 3 for 25 cts.; 12 for 75 cts.

Petunia. Single dwarf, a bushy free-flowering sort, fine in beds and very useful in vases, window boxes, etc.   Price 10 cts.; 3 for 25 cts.; 12 for 75 cts.

RUELLIA.

Ruellia Makoyana (new). This new plant forms a graceful, branching subject, one of the principal attractions of which is a beautiful velvety foliage, with the upper surface of a rich olive-green, shading to purple, relieved with prominent silvery-white midrib and veins. The under side, however, is a bright purplish wine color. This acquisition is already in great favor. It is one of the most brilliant flowering plants known, specimens having been shown measuring thirty inches across, and carrying more than three hundred flowers open at one time. These are of large size, and of a bright carmine red; and not only do they appear in the spring, but also most abundantly in autumn. (See cut.)   Price 30 cts.

PRIMULA (PRIMROSES).

Forbesi (the Baby Primrose). We pronounce this one of the most valuable plants for flower lovers introduced for years, and that it is one of the most profitable plants to grow for cutting. It begins to bloom in thumb pots, when the leaves are scarcely two inches high, and continues to bloom for ten months in succession, the plants soon forming dense clumps of foliage crowded with dozens of long flower spikes. The blossoms are very dainty and graceful, not quite one-half inch across, of a most pleasing rose color, with eye or center of pale gold. They are carried in tiers, on erect but delicate stems, twelve to fifteen inches long, and positively remain fresh for two weeks after being cut. It is unequalled in keeping qualities by any other cut flower that we know of. We must add to all this that the plant grows vigorously and without petting, in shade or in sunshine—even in a cool house—and that many will carry one hundred flower spikes, and more, during the season. It is most remunerative to the trade. (See cut.) Price 20 ⁂s.

Obconica. A lovely perpetual blooming Primrose, needing about same culture as the Chinese varieties. The flowers are borne profusely, in trusses, on long stems, color white, delicately tinted rosy lilac. Of easy culture, thriving with little care and attention. The blooming qualities are wonderful, in full flower from October to August, fully nine months in the year, producing twice the quantity of flowers of any plant with which we are acquainted. Can be planted in open ground in summer and potted again in fall. (See cut.) Price 15 ⁂s.

Chinese Primrose (single). This florists' flower is held in great esteem, and is one of the most desirable winter blooming plants, flowering during the winter and spring months. Choice fringed varieties in many colors. Primroses prefer rather cool treatment and need but little sun, even doing finely in a north window. Price, fine large plants, from 4-inch pots, in bloom, 25 ⁂s.; three varieties for 60 ⁂s. Smaller plants, 15 ⁂s.; two for 25 ⁂s. The large size plants go best by express.

ACALYPHA SANDERI.

Without a doubt the most sensational new plant introduced for many years, and one of the most striking flowering ornamental plants we have ever seen. It is of strong, free growth, with large dark green leaves, from each axil of which, one to two feet long and nearly one inch thick, rope-like spikes of velvety crimson flowers are gracefully suspended. The plant is in flower the year round and is as easily grown, simply requiring a warm temperature to develop its full beauty. Price 50 ⁂s.

CALLA.

Fragrance. This is another of Mr. Luther Burbank's most valuable novelties. We believe that it will become far more popular than the old Calla has ever been, and that it will soon take the old Calla's place altogether, inasmuch as it produces more flowers and requires but half the space. Mr. Burbank says: "In the new Calla, 'Fragrance,' which is one of many thousand seedlings, we have not only a most charming flower produced in a profusion never before surpassed, if equaled, but also with a genuine, sweet, lasting fragrance all its own, but similar to the fragrance of violets and lilies. The plants are of medium size, compact in growth, and multiply with great rapidity, growing and blooming profusely under any ordinary treatment. In purchasing 'Fragrance' no mistake need ever be made, as the foliage is unique, being handsomely fluted." (See cut.) Genuine stock. Price, small, 25 cts.; larger, 40 cts.

VERBENAS.

Verbenas being a leading specialty, we have selected from a large collection the following varieties, which combine the best bedding qualities and embrace the greatest variety of color, which cannot fail to give satisfaction to all. The list includes many of the best so-called mammoth sets; but where older sorts are still the best, we retain them on our list. We also have in stock many kinds not on the list.

Leola, magenta, white eye; *Marion*, mauve, white eye; *Candidissima*, pure white, extra; *Bernice*, pink, dark center; *Beauty of Oxford*, mammoth pink; *Black Hawk*, rich dark maroon; *Climaxer*, extra large, scarlet; *Negro*, nearly black, very fine; *Crimson Bedder*, bright crimson; *Invincible*, purple, large white eye; *Swanger's Beauty*, finest striped sort; *Antonio*, deep blue, very fine; *Glow Worm*, brilliant scarlet, extra; *Gen. Custer*, crimson, white eye; *Mrs. Cleveland*, fine, pure white; *Mrs. J. C. Vaughan*, light pink; *Machadon*, deep blood red; *Miss Arthur*, dazzling scarlet; *Turquoise*, fine blue, white eye; *Loadstar*, large, dark lavender.

Price, customer's selection, 8 cts. each, 75 cts. per doz.; our selection of best sorts, only 6 cts. each, 60 cts. per doz.; fine seedlings, mixed, 50 cts. per doz. Verbenas cannot be sent by mail after May 15. While we shall fill all orders for Verbenas in full as far as possible, we reserve the right to substitute good kinds in place of any of which we may be out.

DRACAENA GODSEFFIANA.

Undoubtedly one of the most striking new ornamental foliage plants of recent introduction. The plant is of an entirely different habit and appearance from all other Dracænas; it is of free-branching habit, and throws out many suckers from the base so as to form beautiful, compact, graceful specimens in a very short time. Its foliage is broadly lanceolate, five to six inches long, and two to three inches wide; of a strong, leathery texture; rich dark green color, densely marked with irregular dots and spots of a creamy white. It is a plant which will undoubtedly be grown very extensively for decorative purposes when it becomes more plentiful, as it is very hardy as a house plant, equaling if not surpassing in this respect the Aspidistra and Ficus.

Price 60 cts.

ALTHEA.

Joan of Arc (new). Or double flowering Rose of Sharon. It begins to bloom almost as soon as planted, and every leaf brings forth pure white double flowers, frequently nine inches round, which are mistaken by many for glorious Camellias. These chaste flowers are borne during the entire season, and even during winter, if the plant is taken up or potted, thereby ornamenting both the garden and the house. As a pot plant, it is most attractive and desirable, and as a shrub in open ground it is hardy as other varieties, and the finest flower yet seen. (See cut.)   Price 30 cts.

ARAUCARIA.

Excelsa (Norfolk Island Pine). As a decorative plant this is one of the handsomest and most serviceable. Its deep green, feathery foliage arranged in whorls, rising one above the other at regular distances, forms a plant of rare grace and beauty. This is also called the Monkey Puzzle Tree. These cannot be sent by mail.   Price $1.00, $1.50 and $2.50 each.

TROPAEOLUM PHOEBE.

A most attractive and deliciously scented variety, producing flowers throughout the winter and summer season, of a deep golden yellow, with a rich crimson-feathered blotch in each segment, outer edges notched or scalloped, the whole forming into an exquisitely shaped blossom. A free, vigorous grower. When in the height of its perfection, strings of growth eight to ten feet long, resplendent with its parti-colored flowers, can be cut for decoration. This new fancy Nasturtium is fine for any purpose where this class of plants can be used.   Price 25 cts.

PENTAS LANCEOLATA.

A pretty half-shrubby greenhouse plant, not unlike a Bouvardia in general habit and appearance, but flowering much more profusely and continuously. It may be planted in the open border during the summer, but it is valuable chiefly as a winter flowering pot plant for the window or conservatory, for which purpose it is especially well suited, being in flower all the time. The pure white flowers are produced in flat

heads of fifteen to thirty flowers each, similar to Bouvardia, but much larger, and last
in perfection a long time. Price 25 &s. each.

PHRYNIUM VARIEGATUM.

Its manner of growth is much like that of a
Canna, the leaves borne on erect footstalks with
spreading blades of oblong lanceolate form. They
are bright pale green, beautifully variegated with
white and gold. In some leaves the whole area is
white, in others only one-half, in others again it is
confined to the interspaces between two or three
of the ribs, while in others it is reduced to a few
stripes. It is very free growing and one of the
most ornamental plants imaginable. As a bedding
plant, planted out in full sun, it is simply elegant,
making a specimen which for ornamental foliage
has no equal. It is a free, vigorous and easy
grower, succeeding in all situations; also fine as a
pot plant. (See cut.) Price 15 &s.

DEUTZIA LEMOINEI.

Without doubt one of the most important new
hardy plants offered in many years. It is a hybrid
between Deutzia Gracilis and Deutzia Parviflora,
and has the advantage over the former of having
flowers nearly three times as large, which are produced in broad based cone-shaped heads of from
twenty to thirty flowers each, and are of purest white, which open out very full. It
is perfectly hardy, and will not only prove a valuable plant for the garden, but is certain to become one of the standard plants for forcing into flower in winter, more especially at Easter, and it is quite sure to take the place of the popular Deutzia Gracilis.
Price, small plants, 20 &s.; very strong, 40 &s.

GERANIUMS.

We call special attention to our entire list of Geraniums. The older or standard
sorts are selections of years of practical growing of this popular plant, while in new
sorts a finer set than the first sixteen in our list cannot be found named in any retail
catalogue in this country. The first seven never having been offered till this year,
being the cream in novelties, from the world's best growers. Our collection is at all
times up to date, which accounts for our immense sale in this class of plants.

Le Fraicheur. Double; white, with a narrow band of rosy pink around each
petal, exactly like a Picotee. Absolutely novel and distinct, and one of the prettiest
flowers we have seen for years; colors delightfully fresh and clear. Price 75 &s.

Dr. Despres. Double; a good truss, composed of large circular flowers; color
changeable; bright violet marked vermilion on the upper petals, the other segments
marked scarlet. Very distinct. Price 35 &s.

Emanuel Arene. Double; one of the most startling novelties of the day. The
floret is large and round, very fine, and of a pure rose color, with immense white eye.
Both colors are very clear and pure, and are combined in the most showy way. Every
florist ought to have this unique novelty. Price 60 &s.

Mme. Rozain. Purest white. M. Rozain, the celebrated raiser, has named this variety for his wife, and pronounces it the superior of all other double whites. Size of floret and truss and freedom all seem perfect. Price 35 cts.

Mme. Hoste. Rich crimson lake at the edge of the petals, shading to rosy carmine towards the center, with startling white blotches on the upper petals; the color scheme is extraordinary, and the size and circular form are of the finest; one of the grandest of the English show type. Extra for house culture. Price, 25 cts.

Vera Vend. Plant of finest habit, large trusses; flowers semi-double, very open, with large petals, the borders of which are rosy orange; large center, striped with white, and delicately marbled and striped with rosy orange-carmine; all the petals bordered with orange-carmine, the center brightened with an aureole of bright orange. A very distinct and charming variety. Price 40 cts.

Francis Perkins. This we consider the finest bright pink bedding Geranium to date for our climate. The plant is a strong, vigorous grower and a prolific bloomer; flowers are perfect in form and are borne well above the foliage on long footstalks; color a clear bright pink. It stands the sun admirably, and for massing there is nothing that equals it in its color. Semi-double. Price 15 cts.

Mme. Jules Chretien (single). Florets full two inches across, perfectly circular and extremely showy. It is the largest, finest and most showy fancy Geranium that we have ever seen. The center of the flower is pure white, surrounded by an aureole of pale violet, and is bordered bright red. The bloom is of the very largest size, and is a beautiful dwarf grower and very free in bloom. A most astonishing variety. Stock very scarce. Price 50 cts.

John Doyle. A strong grower, throwing its truss up well above the foliage, of enormous size, and of the richest bright scarlet. A phenomenal bloomer, as it will produce one-third more flowers than any Geranium we grow. Price 10 cts.

Fleur Poitevine (single). Flowers large, color brilliant rosy carmine, marbled and striped in white, and producing an aureole of deep orange-carmine at the center; the plant is robust and produces quantities of bloom. Price 15 cts.

Md. Jaulin. A new semi-double Geranium; an entirely distinct color. Very large florets compose a truss of grand size; center of flower very delicate pink, bordered with pure white. For freshness and beauty this variety is unrivaled by any bedding sort, and is equally valuable for pot culture. Price 15 cts.

J. Sallier. A vigorous growing and remarkably free-flowering variety. Trusses very large and of perfect form, composed of single flowers; entirely distinct from any existing variety. Edges of petals carmine-lake; the center is washed in tints of rose and bluish heliotrope color, with orange center. Price 10 cts.

H. Dauthenay (single). Low growing, free branching and extremely free in bloom; the trusses are of gigantic size, flowers enormous and quite round, color coppery orange-red, with white eye, with peculiar shadings about the eye; the color is magnificent; the plant blooms in the greatest abundance. Price 10 cts.

Juliet (new semi-double). Its merits are very strong, rugged growth; but short jointed and bushy, in fact a model Geranium, in foliage and habit. Flowering very profusely at all seasons; color pink, with salmon shadings; grand. Price 10 cts.

Dr. Marest. A new dark flowering bedding variety. Wonderful profuse blooming in vases and flower-beds. Price 10 cts.

Deuil de Miribel. The grandest single scarlet that we have yet seen. Footstalks fifteen inches long and very rigid, and bearing enormous trusses of immense circular florets of velvety crimson-scarlet. This variety has the largest floret, finest truss, longest stem and best color among single reds; fine free grower. Price 20 cts.

2

"MARS"

Mars. This new single border Geranium produces more flowers than any other Geranium; though of smaller size, they are produced in such profusion as to completely cover the plant; new color, white and salmon, habit dwarf, leaf with a dark brown zone; equally as good a winter as summer bloomer.(See cut.) Price 15 cts.

Raspail Improved. Among double Geraniums there is nothing finer. The color is a deep scarlet of exquisite shade. The floret is very regular in form, more than semi-double, and about two and one-half inches in diameter. Foliage and habit are both good, and it is a first-class winter bloomer.   Price 20 cts.

Alphonse Ricard. Color bright vermillion-scarlet, habit dwarf, very vigorous, truss very large, measuring from three to six inches in diameter. Large, semi-double floret, sometimes exceeding two inches in diameter. The finest large-flowered Bruant that has ever been introduced, and the best bedding scarlet. It is a constant bloomer from early in the season to late in the fall.   Price 20 cts.

Md. Bruant. A grand new single variety wholly distinct from any other in cultivation. White, veined with carmine-lake, florets regularly bordered with bright solferino, the arrangement of color is exquisite; the trusses are large and beautifully formed; the plant is a good healthy grower and very free-flowering. It rivals many of the fancy Pelargoniums, with its wonderful combinations of colors, and excelling them in blooming qualities, being in flower summer and winter, at all times in bloom.   Price 10 cts.

Mrs. J. M. Garr. Probably the finest of the single white bedders; semi-dwarf, the plants have been a cloud of bloom all summer. Large trusses, and wonderfully floriferous when planted in open ground. Also a fine pot plant. Price 10 cts.

Mme. Ch. Dabouche (double). Extremely free in bloom, with enormous trusses of large round flowers. Color, bright rose, shading to apple bloom pink. A new fancy variety that will attract in any collection; extra good, Price 15 cts.

Mme. Jouis. A very beautiful variety; color, bright rose at the center with a white border of flesh color or white. Florets of the very largest size, semi-double, and flat in form. A grand addition to the doubles; extra. Price 10 cts.

M. Remy Martin (double). Dwarf and free branching, producing flowers abundantly throughout the season. Enormous spherical trusses high above the foliage. Color, beautiful dark rose, marked pure white on the upper petals, lilac on the two lower petals. One of the best of the season. Price 15 cts.

Baron Duranteau. Magnificent double flowers, of a fine shade of crimson, upper petals marked scarlet; greatly admired. Price 10 cts.

F. L. Voith (single). Color a beautiful, clear and brilliant carmine-rose, with a very large and pure white eye; florets round and of perfect form; trusses of enormous size and beautiful shape; plant very free and of excellent habit. Price 10 cts.

Baronne de Scalibert. Dwarf and compact in growth, and so free in bloom that the large trusses cover the plant. Florets large, of bright soft pink color, the centers marked white, petals veined red, a lovely color; single. Price 10 cts.

Mme. Grillet. Of a beautiful, soft China rose color, of a different shade from any other variety that we offer. Very fine double sort. Price 10 cts.

Theodore de Bandville. The best yet produced of the golden type. A good grower and free bloomer, with large well formed florets borne in large trusses; pure yellowish scarlet, semi-double; a very fine sort. Price 15 cts.

P. Crozy. The foliage shows the blending of the two races, having the form of the Ivy and the substance and size of the Zonals. The flowers are brilliant scarlet, borne in the greatest profusion, the trusses measuring six inches across; habit is compact; the flowers are semi-double and last a long time. Price 10 cts.

Souvenir de Mirande (single). It has round florets, upper petals cream-white with a distinct rosy pink border; lower petals salmon-rose, streaked with pure white. A most novel color, and extremely free-flowering. Price 10 cts.

Claire Fromont. Immense sized trusses, semi-double florets of the largest size; growth and habit of plant perfect; color beautiful rose, marked white, without any solferino or magenta shading. The best pink variety we have seen. Stock of this sort very limited. Price 15 cts.

Daybreak. An entirely new color in single Geraniums, being a beautiful light flesh, like Daybreak Carnations; large trusses and grand foliage. We recommend this as one of the finest of the fancy flowering kinds. Price 10 cts.

La Favorite. This eclipses all other double white Geraniums. The trusses are very large and perfect in form, and of the purest snowy white, even when grown in the open ground. A good grower and free-flowering. Price 10 cts.

M. V. Noulens. Very distinct, of the "Mirande" type, clear white with scarlet border, flowers large and good form. Price 10 cts.

Sam Sloan (single). Deep crimson-scarlet; trusses of immense size. Fine for house or pot culture, but its great merit lies in its wonderful blooming qualities when planted out. Price 10 cts.

Mrs. E. G. Hill (single). The center of each petal is a soft, light salmon, bordered with rosy salmon and veined with rose; trusses very large and composed of an immense number of florets; extra. Price 10 cts.

Montesquieu. Very large semi-double florets of exquisite pale mauve, with white eye. It bears a fine truss on a long, stiff footstalk. One of the best. Price 15 cts.

Queen of the Belgians. Single Zonale, pure white; florets and trusses nearer perfection than any other of this class, bringing the whites up in size and form to the colored varieties; free-flowering; fine habits. We recommend this sort for its purity, perfect form and free habits.   Price 10 cts.

Gen. Grant. One of the best single scarlet varieties for bedding, vases, etc. Price, from 2-¼in. pots, 10 cts.; 75 cts. per doz.; 3-in. pots, $1.25 per doz.

S. A. Nutt. A new dark bedding variety, now used extensively in the large parks, and called the best of the dark double varieties.   Price 10 cts.

Beaute Poitevine. Undoubtedly the best double Geranium of its color. Very large semi-double florets of beautiful and distinct form, and borne in immense trusses; very free bloomer, and as a bedding variety it has no equal in its color, brilliant salmon, brighter at the edges, and is a plant of fine short jointed habit, the best of all the salmon shades, either for house culture or bedding.   Price 10 cts.

Mrs. Hayes. Very double, of a beautiful light pink flesh color; trusses of immense size. Sure to please.   Price 10 cts.

Heteranthe (Double Gen. Grant). We consider this the best bedding Geranium of all varieties. The plant redounds in vigor; foliage is strong and striking. Color of the flower is a bright vermillion-red, of pleasing shade. The flowers are absolutely perfect in shape, color and make-up; trusses exceptionally large and borne in immense spherical balls, often measuring eight inches in diameter.   Price 10 cts.

FANCY LEAF GERANIUMS.

Mrs. Parker. In this fine variety recently introduced, we have a combined beauty in flowers and foliage not found in any other Geranium. Its leaves are deep green, with a broad margin of white; flowers perfectly double, bright pink. Splendid for house culture. (See cut.)   Price 15 cts.

Chieftain. One of the best of the bronze type; fine for pots, vases, or bedding; foliage light golden, with a rich chocolate zone. Price 10 cts.

Happy Thought. An entirely new style of leaf, having a large yellow blotch in the center of leaf, with an outer band of green. Price 10 cts.

Mrs. Pollock (golden tricolor). The ground color is a deep green, with a zone of bronze, crimson and scarlet, with margin of clear yellow.
Price 15 cts.

L'Enfer. Plant very dwarf, with small leaves of blackish green and black zone. Flowers fiery scarlet; a novelty in black and scarlet, and entirely distinct among foliage Geraniums, and yet rare.   Price 15 cts.

Mountain of Snow. Pure white, margined; flowers scarlet.   Price 10 cts.

Golden Band. Gold margined; cherry colored flowers.   Price 20 cts.

Cloth of Gold. Fine golden foliage.   Price 20 cts.

Md. Salleroi. A distinct variety, with leaves from one to two inches in diameter; the center of olive-green, with broad margins of white.   Price 10 cts.

Crystal Palace Gem. Golden foliage; center of leaf bright green. One of the very best of the golden varieties; scarlet flowers. Fine for window.   Price 10 cts.

Distinction. Leaves are circular, and very dark green, having a narrow band of jet black near the margin; distinct from anything in cultivation. Price 15 cts.

SCENTED GERANIUMS.

Peppermint. Large velvety leaf, strong grower. Price 10 cts.

Mrs. Taylor. Strong rose fragrance, and deep scarlet flowers. Price 10 cts.

Lemon. Large leaf, scented. Price 10 cts.

Rose. Large leaf, scented. Price 10 cts.

Skeleton Leaved. Rose scented. Price 10 cts.

Quercifolia (oak). Leaves marked with black. Price 10 cts.

Fernæfoleum Odorata. A valuable and scarce sort; its leaves have the appearance of the Fern, also an odor similar to Sweet Fern. Price 15 cts.

Lemon. A dwarf small leaf sort, growing dense and compact. Price 15 cts.

Nutmeg Scented. Small leaves, resembling "Apple Scented." Price 10 cts.

Variegated Rose Scented. Foliage green and white. Price 15 cts.

Birch Leaf. Small leaf, with birch fragrance. Price 10 cts.

Attar of Roses. Leaf apple green, large and velvety. Price 15 cts.

Shrubland Pet. Small light green leaves, very fragrant. Price 10 cts.

IVY-LEAFED GERANIUMS.

Gen. Championnet. Flowers double and of enormous size, equally as large and double as the finest Zonale; color, deep scarlet with carmine shadings. The largest flowered of all the Ivy Geraniums. Price 15 cts.

Mme. A. Guillemand. Flowers full and double; color, violet-purple; extremely grand of this color. Price 10 cts.

Souv. Chas. Turner. The handsomest double Ivy Geranium ever introduced. Florets two and one-half inches, in trusses six inches across. The color is a bright pink, approaching scarlet; the upper petals, feathered maroon. Price 15 cts.

Joan of Arc. This is one the most charming plants we have ever offered. The flowers are perfectly double, white as snow, and literally stud the plant when in full bloom. The foliage is extremely handsome, the dense, glossy green leaves making a most effective background for clusters of ivory-white flowers. Price 10 cts.

L'Elegante. An extremely pretty variety, of bright green foliage, with broad band of creamy white, often margined with pink; its pure white, single blossoms are produced in dense clusters. Price 15 cts.

BEGONIAS.

· In no other class of plants has there been so much improvement, so many additions in choice varieties, and so much progress in popularity during the past few years, as in the Begonia. The reasons are: ease of culture, rapid growth, freedom from insects, fine flowers, great variety and wonderful beauty of foliage, and all that goes to make the window garden enjoyable, can be found with Begonias alone. Our collection is one of the finest in the United States. No expense spared to get the best from the world's most noted growers. The coloring and form of varieties offered are very unique and beautiful in the extreme.

Glory of Lorraine (new). The greatest novelty in Begonias of the year. The plant presents a mass of floral loveliness for nearly the whole year—blooming autumn, winter and spring—a perfect bouquet of clear pink flowers. Price 50 cts.

Double Vernon (new). Like "Vernon," but with double flowers; will be used in large quantities for bedding, vases, etc., soon as florists can supply it at moderate cost; flowers last much longer than the single variety. Price 25 cts.

Masterpiece (Rex). Exceedingly highly colored, leaves having a pinkish metallic lustre, entire center being practically pink, with a very narrow outer edge of bronze-green. It is as highly colored as a sea conch, having a bright pinkish metallic lustre impossible to describe; exceedingly beautiful.   Price 30 cts.

Bertha McGregor. A splendid new seedling and one of the most beautiful fancy Begonias yet seen. A cross between the Rex and flowering section, showing a combination of coloring and great freedom of growth, together with fine habit, which will make it of great value to both amateur and florist. It grows readily under ordinary treatment. Leaf six by nine inches, long pointed and with six deep notches, producing foliage in abundance; the center of leaf is small, dark and palm-shaped; the body of leaf being solid silver, outlined with bronze; elegant. (See cut.)   Price 25 cts.

Pictaviense (Scharffiana X Metallica). This cross has produced a plant of fine habit, fine foliage and fine flower. The leaves are intermediate between the two parents, both in size and form; the under side is a rich purplish red, the veining very prominent, and the face of the leaf a fine bronzy green.   Price 15 cts.

Mme. Lionnet. The ground color of the leaf is a rosy bronze, distinctly overlaid with a silvery-metallic lustre, the entire surface covered with crimson pile; the best red-leaf Begonia on the list, being very brilliant in color. The flowers are bright pink.   Price 25 cts.

Pres. Carnot (Rex). Very beautiful; strong, vigorous grower; light brown foliage beautifully marked, giving it the appearance of frosted silver over the larger part of the leaf. One of the most beautiful of the newer introductions. Price 25 cts.

Speculata. Quite a novelty in Rex type. Leaves are in the form of a grape leaf. Color a bright green, with background of chocolate; veins of light pea-green, the whole leaf spotted with silver. In bloom it is magnificent, the panicales composed of great numbers of individual pink blooms, are lifted high and spray-like, quite clear of the foliage. (See cut.)   Price 15 cts.

Countess Louise Erdody. The leaf, which has a metallic lustre, appears dark silvery in the center, shading into coppery rose toward the margin, which is broadly and evenly edged with the same hue, but darker and more brilliant. The veins are yellowish green on both sides. The striking peculiarity, however, which distinguishes it from all other Begonias, consists in the fact that the two lobes at the base of the leaf do not grow side by side, but one of

them winds in a spiral-like way until in a full grown leaf there are four of these twists lying on the top of the leaf nearly two inches high. (See cut.) Price 15 cts.

Metallica. A variety with a peculiar metallic-like lustre, and charming rose-colored flowers; very worthy variety, combining both beauty of leaf and flower in same plant. Winter flowering; extra. Price, 10 to 25 cts.

Revolution. A very striking new variety, with a very pronounced double whorl on the leaf, one above the other; the texture is fine and velvety and the shading is beautiful, showing a nice silver zone. A free, thrifty grower. A fine additon to the whorled Rex. Similar to "Countess Louise Erdody" in shape of leaf, but entirely distinct in its coloring and a more sturdy grower. Price 25 cts.

Mme. Leboncy (Rex). A very rare sort, and none in our list can equal it in rich. lustrous coloring, which it is impossible to do justice in a description. The young leaves are reddish metallic, studded with irregular silvery spots. The older leaves are dark silver, shaded steel and spotted dark green; the entire leaf shaded red and burnished as in tinsel. Price 40 cts.

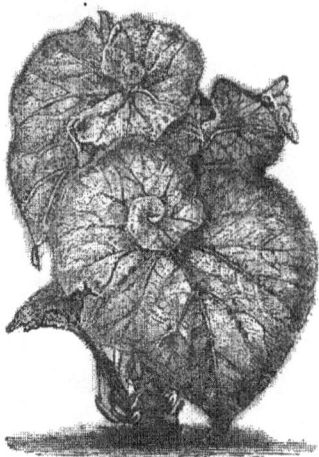

Argentea Guttata. This variety has purple-bronze leaves, oblong in shape. with silvery markings, and is in every way a most beautiful Begonia. It produces white flowers in bunches on ends of growth stems. (See cut.) Price 15 cts.

Tuberous-rooted. These splendid Begonias are very beautiful, and unsurpassed for bedding or pot culture. The plants present a striking appearance, being covered with magnificent showy flowers. The bulbs or tubers can be easily grown, requiring like Gloxinias, light loamy soil and a somewhat shaded situation. After blooming all summer, the bulbs, if in beds, should be taken up in November. dried off like Gladiolus, and packed in a box of coarse sand placed in a dry place, secure from cold. If in pots, they can be gradually dried off in November by withholding water; when well dried out, put the pots containing bulbs in a dry warm place, until spring, when the roots should be shaken out and re-potted. We can furnish the following colors : red, white, pink and yellow. All strong flowering bulbs. Price 20 cts. each.

Manicata Aurea (from Italy). A very beautiful and distinct ornamental Begonia, with clear and glossy green foliage, marked and blotched with cream color, deepening to bright canary; flowers pink, lace-like, in long spreading panicles. Its heavy glossy leaves being boldly blotched with a rich golden cream, and clear carmine etch-ing in the matured leaves, altogether making the handsomest variegated plant known.   Price 25 cts.

Rubra. A fine acquisition to our winter flowering plants. The leaves are of the darkest green; color of flowers, scarlet-rose, glossy and wax-like.   Price 10 cts.

Souv. de F. Gaulian. A remarkably strong growing variety, of stiff, upright habit; foliage very large; flowers beautiful coral red, in large pendant panicles simi-lar to "Rubra," but very much larger and finer.   Price 15 cts.

Louise Closson. One of the richest and most beautiful of the Rex family, with large and very bright foliage. The coloring is simply magnificent; center of leaf deep bronze with broad band of silvery rose; edged bronze, spotted rose. (See cut.) Price 25 cts.

Queen Victoria (Rex). Solid silver leaf of crepe-like texture, red veins, and fluted edge; plant a strong grower.   Price 25 cts,

Incarnata. A splendid winter flowering plant. About the Christ-mas holidays this is covered with one mass, from top to bottom, with bright pink flowers, looking like one huge bouquet. The Christmas flowering plant for everyone.
Price 10 cts.

Vernon. One of the very best summer or winter blooming Begonias. Flowers when small, and is literally covered with flowers as the plants attain age and strength. Numerous flowers are of a brilliant orange-carmine color, and the foliage a glossy red which grows more intense with the advancing season.   Price 10 cts.

Clementina. A cross between Rex and "Diadema." The color of the stem and upper surface of the leaf is a beautiful bronze green, ornamented with large silver spots, arranged parallel with the ribs of the leaf; the under side is a bronzy red, pro-ducing a beautiful effect. An easy and rapid grower.   Price 20 cts.

Weltoniensis. Beautiful old-fashioned pink flowering variety; also known as Coral Begonia.   Price 10 cts.

Thurstonii. Bright red foliage and veinings underneath the leaves, and bright metallic green, shading to red in the younger growth on top, with the deep veinings; the flowers are a beautiful deep pink in bud, but when expanded become a beautiful shell-pink. A grand improvement on "Metallica."   Price 15 cts.

Multiflora. Winter flowering; constantly in bloom; a profusion of rosy pink blossoms from November to April.   Price 10 cts.

Riciniflora. A magnificent decorative plant; leaves of immense size and similar in form to "Ricinus." Flowers in very large panicles on flower stalks two to three and one-half feet high; light pink.   Price 15 cts.

McBethii. A beautiful white or winter flowering sort, of dense and finely cut foliage, blooming profusely through the winter months. Price 10 ćts.

Pond Lily. Named on account of resemblance of leaf to that of the Pond Lily; fine foliage; one of the best winter blooming varieties; flowers pink. Price 15 ćts.

Roi Ferd Major (Rex). Large silver leaf, center blotched dark green, edge ruffled, and covered with blotches of dark green with red shading. Price 25 ćts.

Silver Queen. Not a new variety, but one that came to us with above name, and we find it one of the finest of Rex Begonias. The entire leaf is a light silvery color, veined green; flowers white, shaded pink. * Price 15 ćts.

Queen of Hanover. Leaf of a soft velvety texture, covered with red pile, center and edge of soft green velvet, the zone formed by small silver dots. Price 25 ćts.

Sanguinea. Leaves dark, glossy olive-green, under side deep red, flowers white; easily grown. Price 15 ćts.

Maple Leaf. The name indicates shape of leaf, which is quite large and of a pure, light, solid green, thick in texture, and surface very glossy. This makes an elegant house plant, and easily makes a fine specimen. Price 20 ćts.

Rex Varieties. A magnificent class of house plants, remarkable for the variety and beautiful markings of the foliage; well adapted to vases and baskets in shady situations; also very fine ornamental plants for window, but should be kept where it is warm and not too sunny. Price, older sorts, small plants, 15 ćts.; large, 25 to 50 ćts.; new and rare sorts, small plants, 25 cts each.

Otto Hacker. Plant, a strong vigorous grower, of stiff, upright habit, with large, shining deep green leaves, eight to ten inches long. The flowers are borne in immense pendant clusters eight to ten inches across, twelve to twenty-five large flowers in a cluster. Color a beautiful bright coral red. Price 20 ćts.

ABUTILON.

Savitzii. This unique maple-leaved variety is from Japan. It is a decided improvement upon "Souvenir de Bonn," inasmuch as the variegation is entirely different, the contrast between the green and white is sharper, and the habit of the plant is dwarfer. It will be of exceptional value for edged sub-tropical beds,—also among the best as a pot plant. Price 25 ćts.

Mrs. G. Laing (new). Immense flowers of bright rosy pink, of beautiful expanded form; habit good, and free in growth and bloom. Flowers nice for cutting, as the stems are long. A great improvement on any variety similar in color. (See cut.) Price 20 ćts.

Souv. de Bonn. The bright green leaves are distinctly edged with a broad pure white band. The flowers are of golden yellow color, veined with scarlet. Either bedded out or as a pot plant, makes fine specimens in a short time.
Price 10 ćts.

Golden Fleece (or Golden Bells). A fine yellow flowering Abutilon of strong vigorous habit and a very free bloomer. Color, rich golden yellow; flowers of large size. This variety blooms the entire year; one of the best for summer bedding or winter window plants. Price 10 ćts.

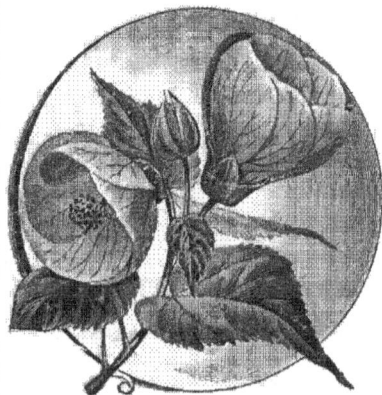

3

Infanta Eulalia. Unlike the ordinary Abutilon of scraggy growth, it is very compact and short-jointed, making a very neat pot plant. The flowers are the most beautiful we have seen among Abutilons, being very large, yet short and beautifully cupped, and of the most lovely pale satin-pink color imaginable. It flowers in profusion, both summer and winter.   Price 20 cts.

L'Africain. In contrast with above we offer this fine new and attractive dark sort. Its color, dark crimson, shading still darker at center; petals very thick and heavy and having a beautiful glossy appearance not found in the other varieties. Price 20 cts.

Boule de Neige. This is one of the best Abutilons yet introduced, with dark green leaves and pure white flowers; it blooms freely, either as a bedding-out plant or pot plant for winter.   Price 10 cts.

Firefly. By far the highest and brightest color of all the family, and one of the freest bloomers. When grown in a pot it flowers all winter, and summer when planted out; color nearest approach to scarlet of any yet introduced.   Price 10 cts.

Thompsoni-plena. A double Abutilon had never been produced until this was obtained. Foliage beautifully mottled yellow and green; perfectly double flowers; color, rich deep orange, shaded and streaked with crimson.   Price 10 cts.

PELARGONIUMS.

This grand old plant is again coming to the front in popular favor. The reason is evident, as seen in the improved sorts of today. The large, fluffy, crimped, and in some sorts semi-double flowers, their longer season of bloom, greater variety of coloring, from the most delicate to the brightest and most fiery. Also in their favor—that many of these improved sorts are much stronger growers than the old varieties. No American florist will catalogue a finer set than the following:

Sandiford's Surprise. A charming flower of splendid habit, big black blotches in upper petals, edged fiery red, surrounded with a broad band of white; lower petals white, bright red spot in center of each; continues long in flower, and is one of the most distinct in cultivation.   Price 40 cts.

Sandiford's Best. A new and distinct flower, of a beautiful shade of pink, edged with a deep band of the purest white; large white throat. The plant is marvelously free-flowering, and the trusses large and round; very attractive and a most charming variety.   Price 40 cts.

Sandiford's Wonder. Splendid semi-double white flower, very pure in color, some flowers showing a small maroon spot in upper petals; heavily fringed; a great advance in this class of plants. The many admirers of the genuine Florist Pelargonium will welcome this grand variety as being simply invaluable for all kinds of floral work, as the plants are smothered in bloom all through the season. Price 50 cts.

Dorothy. Rosy salmon, dark maroon blotch on upper petal, richly shaded plum color round the throat, petals prettily fringed, give a light, elegant appearance to the flower; begins to flower early in the season, keeping covered in bloom to the end.   Price 30 cts.

Champion. Fine large flowers, blush white, with maroon-crimson blotch on upper petals, splendid habit, making a dense, bushy plant smothered with bloom; the finest of all the spotted varieties.   Price 25 cts.

Countess. Bright cerise, with dark blotch on upper petals and white throat, the last being pure and well defined; gives the flower a striking appearance; plant very dwarf and free.   Price 25 cts.

Edward Perkins. Bright orange-scarlet, dark maroon blotch on upper petals; beautifully fringed and undulated; good habit, very early and free flowering.
Price 25 cts.

H. M. Stanley. Bright, rosy crimson, upper petals shaded scarlet and heavily blotched; the flowers are produced in marvelous profusion.   Price 25 cts.

Marie Malet. White, with maroon-crimson blotches on all five petals; remarkably dwarf and free flowering.   Price 25 cts.

Goldmine. Bright orange, white throat, feathered in upper petals; habit good; a grand sort.   Price 25 cts.

Mrs. Robt. Sandiford. Grand white variety, flower large size, well doubled, beautifully ruffled along the edge of the petals, and of glistening snow white, called doubled white from the crimped and fluffy appearance of the flower, which has also an extra petal, allowing it readily to pass for a double white bloom; a strong, vigorous grower.   Price 25 cts.

Madam Thibaut. White, richly blotched and marbled with rose, the upper petals marked with crimson-maroon; large white center, immense trusses of large, fine flowers with undulated petals. A strong grower.   Price 25 cts.

Apple Blossom. One of the finest blooming house plants in our entire list. People never tire of this, the freest flowering of all the Pelargoniums. Flowers, shades of pink and white, and slightly frilled. It flowers from eight to ten months of the year, and when at its best, the flowers are so profuse they nearly hide the foliage, it being one mass of pink and white.   Price 20 cts.

Victor. Color of upper petals almost a black, and very velvety; lower ones a bright lively crimson; center pure white, covering nearly one-half of bloom; florets extremely large and very showy. The finest of its class. The great market sort, tens of thousands raised for the New York market alone, meeting with ready sale.
Price 20 cts.

Belle Blonde. Color, clear rose; large maroon blotch on the upper petals; a very free bloomer; strong growth, in fact, grown as easily as a Geranium, and when in full bloom it makes one of the grandest and most beautiful specimens among the whole family.   Price 15 cts.

CARNATIONS.

The rich spicy odor of the Carnation, combined with its varied colors, handsome form, leaves but little to be desired. They are very easily grown, and bloom freely as bedding plants in summer, or window garden in winter. Planted out in April they will commence flowering in early summer, and continue until checked by heavy frosts in autumn. If intended for winter flowers they should be gone over every three weeks, and the young growth cut back to within four of five inches of the main stem. After August 1st they should be allowed to grow and bud, as by September 15th they should be taken up and potted. Carnations are a specialty with us, and we very much doubt if our customers can find as select a list published in any. florist's catalogue in the United States. Only the very best are allowed in our collection.

Bon Ton (new). Bright, warm scarlet; large, deeply fringed, fragrant flower, on a stem as stiff as a reed; perfect calyx; habit of the plant exceptionally strong. The growth is so strong that it does not need staking. Comes into bloom with a fine crop for Christmas, when scarlet Carnations are so much in demand, and increases in quantity and quality as the season advances. A great favorite in the Boston market, where it brings the highest prices.   Price 15 cts.

White Cloud (new). It commences blooming among the earliest, and throws a fine, long, stiff stem from the begining and continues right through the season. Flower large, excellent form and finish, good substance and very full. Strong, fruity fragrance. Calyx strong and holds the full bloom well. We consider this one of the best whites we have ever grown. Certificated, and a prize winner wherever shown.
        Price 15 cts.

Conch Shell. This new variety might have been called the pink of many shades, as from a bed in our house we can cut flowers in shades from quite a bright pink to the palest tint of pink. A free blooming sort, quite dwarf in growth; fine for growing in pots, or summer blooming in the open ground, and one of the best for florists.   Price 15 cts.

Gold Nugget. The nearest to a pure, deep yellow of any sort yet introduced. Strong growth, free blooming from early to late; long, strong stems, large flowers of pleasing form, and not bursting; center petals standing erect, apparently diminishing the actual circumference.   Price 15 cts.

Psyche. White, flaked scarlet; flower of extra size, but not crowded with petals; long, stiff stem; a very profuse bloomer. This is a great producer of flowers: in fact, it promises, on account of its great freedom of bloom, to be a desirable variety from the money or commercial point of view.   Price 15 cts.

Wellesley (new). The brightest, most brilliant scarlet of any sort we have ever grown. Bloom of medium size, on stiff stems; a fine business pink; its free blooming qualities, fiery color, will make it a favorite with both the grower and buyer of cut flowers.   Price 10 cts.

Mrs. S. A. Northway (new). It is a profuse bloomer bearing very large flowers on exceedingly strong, stiff stems, blooms are of the most exquisite form, very full and double, the center is high built, the edges of the petals are beautifully serated, the calyx never bursting. This variety possesses a most delicious fragrance that is lacking in so many of the newer sorts. When first opening, the flower is white, lightly shaded with a lovely tint of soft shell-pink, but as the flower becomes fully developed it changes to almost pure white.   Price 10 cts.

Empress. One of the grandest of recent introduction; blooms of immense size, on strong stems; color, the finest dark crimson, and an improvememt on all varieties of its color. Only Flora Hill and G. M. Bradt equal this in size, and it's away ahead of all others in list in strong growth and heavy foliage.   Price 15 cts.

Argyle (new). Color, a beautiful carmine, unlike, in color, any other sort in our list. Strong grower, flowers very large; a fine fancy pink.   Price 15 cts.

Painted Lady (new). A large, very full flower, with perfect calyx and extra stiff stems. A model grower with firm curling grass; color, brilliant cherry-pink, at times laced lighter pink about the edge; one of the most profuse bloomers in the family. We believe this variety possesses all the requirements of a good Carnation, being very free in bloom; a variety that must be seen to be appreciated. Price 15 cts.

Flora Hill (new). The grandest white variety to date, and destined to be found in every collection, because of its easy management. The flower is of enormous size, and rounded built; wonderfully free in bloom, stems strong enough to support the blooms nicely; good calyx; not subject to rust. Without exaggeration we can claim this as the most prolific bloomer in the Carnation family. All Carnation lovers everywhere must give this wonderful variety a trial. (See cut.) Price 10 cts.

Jubilee. The red Carnation for which everybody has been looking. Color, intense scarlet, of the richest shade. One of the largest flowers among reds. Calyx very strong, never bursts; flower quite full and well built. A persistent bloomer, and very free. Price 10 cts.

Geo. M. Bradt. Color, clear white, heavily edged and striped with bright scarlet, giving it a very bright and cheerful appearance. Flowers large and full, with center petals standing erect, giving the bloom a well-rounded form. Habit strong and vigorous. This variety has proven the most even and constant bloomer from early to late. Received gold medal at Atlanta. Certificate at same place and at Chicago. One of the largest in our collection. Price 15 cts.

Eldorado. A strong vigorous plant, healthy foliage, and finely formed flowers. Light yellow, or perhaps, nearer a buff, and free from all white markings. Petals edged with narrow band of light pink, almost a picotee in marking. Price 10 cts.

Armazinda. Pure white, lightly penciled with scarlet. Fine large flower of good form, on very stiff stems nearly two feet in length. Calyx very firm, and flower does not burst. Very fragrant. Plant a healthy strong grower and free bloomer. One of our best.   Price 10 cts.

Tidal Wave. Flowers very large and very perfect in form; color, bright rosy pink, changing to a beautiful soft pink with salmon shading, when flowers are fully expanded. A healthy grower, and free bloomer. Plant dwarf.   Price 10 cts.

William Scott. Clear pink, early and free; the blooms are perfect and good size. A great improvement over all the bright pink sorts; a continuous bloomer. Sure to please. A great market sort.   Price 10 cts.

Lizzie McGowen. Flowers pure white, with serrated petals, keeping from ten to fifteen days after cutting; is borne on long stiff stems; does not burst. Price 10 cts.

Daybreak. It is of the most perfect habit, flowers of the largest size, very double and full in the center, the center petals are slightly raised; the petals are thick and of heavy texture, and the flowers keep perfect a long time after cutting. In color it is a beautiful and delicate pure bright flesh color with no salmon shading, and is entirely distinct from any other variety now in cultivation. Finest as a winter bloomer; also the best of all Carnations for bedding and summer bloom.   Price 10 cts.

Annie Webb. Fine dark crimson; grown largely by florists for cut flowers, and considered one the best.   Price 10 cts.

Mrs. Fisher. At this time the finest white Carnation sold in the Boston market in quantity are of this sort. Nearly all the flowers are pure white, but at times in midwinter it shows a slight pink tint. This is fine for winter, and also the best white for flowering in open ground in summer.   Price 10 cts.

PALMS.

Cocos Weddeliana. This beautiful Palm is unquestionably the most elegant and graceful in cultivation. It is admirably adapted for the centers of jardinieres and fern dishes, as it retains its freshness for a long time, while for dinner-table decoration it is unexcelled, and should be in every collection. (See cut.) Price, young plants, 40 cts.; larger, 75 cts. and $1.25.

Latania Borbonica. This is the typical Palm, and is more largely used than any other. Its strong, healthy habit commends it to all, and gives it a fitness for window and room culture not possessed in the same degree perhaps by any other; certainly among the best. Price 30 cts., 60 cts., $1.25 and $3.50 each.

Phœnix Reclinata (Date Palm). Beautiful reclinate foliage of graceful habit. Easily grown, and one of most attractive and ornamental sorts in our list. Price 60 cts.; specimens, $4.00.

Kentia. The beautiful "Thatch Palm." The petioles are a bright green and the leaf divisions narrow, very delicate and graceful. Handsome for table or any kind of decoration; one of the best on our list; very ornamental. Price 50 cts. and 75 cts.; specimens, $1.25 and $2.25.

Areca Lutescens. One of the most valuable and beautiful Palms. Its dark glossy green leaves are gracefully curved on slender stems, and the entire foliage is gracefully disposed. The trunk and stems are golden yellow.

Price, fine specimens, $1.25 and $2.50.

Corypha Australis. A fine variety. Shape entirely distinct from other sorts.

Price, fine plants from 4-inch pots, 60 &s.

Cycas Revoluta. Known as "Sago Palm." Price, good plants, $1.50 to $3.

NOTE.—Palms do not show their character leaves till from one to two years old, and some even later. Therefore, in buying small plants, do not be disappointed with shape of leaf, as it will be the seed leaf, and the perfect, developed leaves will come later. The larger sizes all have full character leaves.

ASPARAGUS.

Sprengeri. A most desirable new species, especially useful to grow as a pot plant for decorative purposes or for planting in suspended baskets; the fronds are long trailing, and of a rich shade of green, most useful for cutting, retaining their freshness after being cut for weeks. It will make an excellent house plant, as it withstands dry atmosphere, and will succeed in almost any position. No introduction of recent years has made such a favorable impression upon us as this Asparagus. (See cut.)

Price 15 &s.; strong plants, 30 &s.

Plumosus. A climbing plant with bright green, gracefully arched foliage surpassing Maidenhair Ferns in grace, delicacy of texture and richness of color. The fronds are twelve to fifteen inches in length, and taper to a point from a width of twelve inches. One of the most beautiful decorative plants. Price 25 &s.

Tenuissimus. Fine foliage. A handsome climbing plant for the window and a very useful pot plant. This and above are also called Lace and Asparagus Ferns.

Price 15 &s.

IRIS.
RARE HARDY SORTS.

Aurea (The Golden Fleur de Lis). This is one of the finest of this beautiful class of plants, and the best of hardy perennials. It grows to a height of four to five feet, bearing twelve to fifteen pure golden yellow flowers of large size, a color which is rare in Iris. A grand plant that should be in every garden. Price 30 &s.

Ochroleuca. A grand Iris, with long leathery foliage, and flowers of the purest white, with yellow blotch on the lip. Flowers very freely in the early summer. A grand plant, rare in gardens as yet, and one that should be in every collection of hardy plants.   Price 30 &s.

Pseudo Acorus Variegata. A fine hardy plant with deep green foliage, broadly striped creamy white. Flowers large; bright yellow, slightly veined brown. A valuable plant for the edges of ponds and streams as it delights in plenty of moisture, but will thrive in any soil.   Price 20 cts.

Kæmpferii (Japanese Iris). These are among the most beautiful of hardy plants and cannot fail to please the most fastidious. Their peculiar and quaint markings can be better imagined than described. They are exceedingly showy and last a long time in bloom. Some are self-colored, while others are marbled and tinted with different shades. The race is perfectly hardy and very easily grown. They require a good rich soil, and if it be possible to plant them near a pond or in a damp position they will thrive the better, though this is not absolutely necessary, as they will make a splendid growth and bloom abundantly under the same conditions as most other hardy plants. Several finest sorts.   Price, each, 20 cts.

ACERATUM.

Princess Pauline (new). A most distinct and novel variety of dwarf, compact habit, rarely exceeding five inches in height, its peculiarity being that both colors, blue and white, are combined in the same flower, the body of the flower being white while the stamens are of a light sky blue, giving the flower a pretty variegated appearance; one of the most distinct new bedding plants that has come under our observation this season.   Price 10 cts.

White Cap. A fine compact bedding sort, with flowers of purest white; also fine for vases or as a pot plant; a very profuse bloomer; the best of all whites. Price 10 cts.

Tapis Blue. Growth very compact and dwarf, blooming in greatest profusion, in fact, a solid bouquet of blue.

Price 10 cts.

Le Geant (new). During late years we have had so many of the charming dwarf Ageratum that are so well adapted for borders and carpet bedding that it is with pleasure that we introduce to the trade this variety, remarkable for its tall growth, making it a valuable plant for the center of vases and large baskets. The branches are red, straight-growing, tipped with very numerous clusters of flowers of the most beautiful blue.   Price 25 cts.

PAEONIES.

The growing demand for these very hardy and showy perennials has led us to increase our stock and also to make up an assortment of the best of each color. They should be planted in deep, rich soil, after which they require less attention than any other class of flowering plant.

Tenuifolia. The double form Tenuifolia is interesting and handsome, not only on account of its rich crimson flowers, which resemble in color the Gen. Jacqueminot Rose, but the fine feathery subdivided foliage makes them a handsome ornamental plant, even if they never produced a flower. This variety is always scarce, and we are now able to offer it for the first time. (Set cut.) Price 50 cts.

Amabilis Grandiflora. White, very large, double, fringed petals, fine, very sweet.
Price 35 cts.

Duchesse de Nemours. Rose pink, very large double, sweet; one of the best; very showy. Price 30 cts.

Elegans. Outside petals dark pink; large salmon center; loose, fine, sweet; an elegant sort. Price 30 cts.

Francis Ortegal. Dark purple crimson; very large, fine, deep, double and sweet. Price 25 cts.

Officinalis Rubra Plena. Rich, deep crimson; very early, and one of the brightest of all dark-colored varieties. Price 25 cts.

Amabilis Speciosa. Rose, fading to pink; large, very double, sweet, high-built center, very free flowering. Price 25 cts.

BOUGAINVILLEA GLABRA.

Sanderiana (new). It affords us great pleasure to offer this new and useful flowering plant; small and large plants alike being covered with blossoms. Even plants in thumb-pots, were laden with bloom. The house in which our plants have been cultivated has been a sheet of flowers from May until December, the plants carrying large numbers of brilliant blossoms. Many excellent judges state that this will be a favorite market plant, as it produces as freely as a Fuchsia its dazzling rosy crimson flowers, lasting so long in perfection, and produced in smallest pots, under all conditions in amazing profusion. A first-class certificate was awarded Messrs. Sander & Co., by the Royal Horticultural Society for this new plant. We have but recently placed this plant in stock; above description is from the introducers, Sander & Co., of England. (See cut.) Price 30 cts.; larger, 60 cts.

4

FUCHSIAS.

Little Beauty. Said to have been named thus, on account of every one upon first seeing it, saying, "isn't it a little beauty." The introducer says he has often counted 150 buds and blooms on one plant in a four-inch pot. Flowers single, about one and one-half inches long; sepals bright red, corolla purple; a fine bushy grower. (See cut.)
Price 15 cts.

Sapley Freres. Erect grower; very large flower; corolla very double of rosy violet; long, recurving sepals of bright coral-red.   Price 10 cts.

Autumn Leaves. This remarkable fancy variety is said to have been brought from across the water by a sailor. We consider it by far the most distinct and beautiful of its class. It is impossible for one not familiar with our eastern forests in autumn, to imagine the beauty of its foliage. In habit it droops, and is most charming when staked and the branches allowed to fall naturally and gracefully down, forming streamers of green, red, bronze and gold, terminating with its beautiful clusters of flowers. Color, tube and sepal, rosy scarlet, corolla, deep violet-crimson. Flowers and foliage blending in perfect harmony.   Price 20 cts.

Mme. Bruant (new). A vigorous grower, and of fine, drooping tree habit. The flowers are of a size and fullness before unknown; the color is a rosy heliotrope, marked and veined in rose. Sepals bright red, and to which a number of the petals of the corolla are very curiously affixed, owing to the extreme doubleness of the flower. Grand, long-pointed, rounded bloom, with sepals strongly recurved. This fine new sort came from France.   Price 25 cts.

Dr. Topinard. This is one of the finest Fuchsias of recent introduction. Sepals large, rich cherry red; corolla large, single, white. One of the best.   Price 10 cts.

Wave of Life. Flowers single, dark purple, golden foliage; the contrast of color in flower and foliage showing to great advantage.   Price 10 cts.

Trailing Queen. This Fuchsia is an entirely new departure, differing from all other large flowering varieties in its habit of growth, as it is a trailing vine. The young plants, as they commence growth, creep out to the edge of the pot, and then go trailing downward. The flowers are borne in large, drooping clusters, and are very large, long and graceful; buds, tubes and sepals being bright rosy scarlet; corolla, when it first opens, a deep, rich violet-purple, changing to a fine shade of crimson.   Price 10 cts.

Mrs. E. G. Hill. This is one of the grandest Fuchsias we have ever grown. Unlike most of the double white Fuchsias, it is a robust, upright grower, not coarse, but compact and very symetrical. The tube and sepals are a bright, reddish crimson; corolla, pure white, extra large and double.   Price 10 cts.

Black Prince. Tube and sepals bright waxy carmine color; sepals with pale green tips and large open pale pink corolla, margined with deep rose.   Price 10 cts.

Storm King. This fine Fuchsia is of German origin, of elegant, graceful habit, producing its immense, finely formed double flowers in great profusion; color, tube and sepal red; corolla, large double white, beautifully tinted with a peculiar shade of rosy pink.   Price 10 cts.

Sunray. Color of foliage as rich and clear as any tricolor Geranium. The leaves are of a bright crimson, white and bronzy green; corolla, purple, sepals coral-red.   Price 15 cts.

Speciosa. A well known variety, producing large flowers two inches in length, tube and sepals of which are blush, the corolla, crimson. Some plants of this variety, grown in eight or nine-inch pots, will produce from three hundred to five hundred flowers from December to May.   Price 10 cts.

Mrs. Marshall. Flower tube and sepals pure white; corolla, carmine; a very fine and profuse winter bloomer.   Price 10 cts.

Ernest Renan. One of the finest on our list. Tube short; white sepals, red corolla, pure rose color; of great value on account of its early blooming qualities.   Price 10 cts.

CANNAS.

The Cannas are now among the most popular of bedding plants. With the newer sorts now being introduced they will be even more in favor.

McKinley. Brilliant crimson with scarlet shading. Trusses large and compact, of good shape, held well above the foliage, and the flower spikes are produced freely and evenly. Dwarf compact habit, growing about two and one-half to three feet high.   Price 30 cts.

Duchess of Marlborough. Absolutely pure pink; grows four to five feet high; leaves green, three feet long, one foot wide; numerous spikes, each with three to four trusses of flowers; each truss bears from twenty to fifty flowers; flowers fully five inches across; trusses frequently are a foot in length.
                                                  Price 25 cts.

Duke of Marlborough. Dark, velvety maroon. The darkest of all Cannas; extremely rich crimson maroon, shading towards purple and black. Fine, erect, compact truss; extremely floriferous. Foliage, bluish, metallic green; height, three or four feet.   Price 30 cts.

Champion. This is absolute perfection. The largest, brightest and most beautiful Canna in existence to date. Immense spikes of gigantic flowers with petals two and one-fourth inches wide, of, a pure, dazzling crimson-scarlet; a color not approached in brightness and richness by any other Canna. It is of very vigorous growth with bright green foliage. Stock very limited.   Price 75 cts.

Pres. Cleveland. A genuine Gibraltar among Cannas. Height, three and one-half feet, but the strongest, heaviest grower in the family; very free in producing compact, broad leaves; heavy flower stock, crowned with large truss of broad petaled florets; color, rich orange-scarlet. There is nothing flimsy about this variety; the mass of color produced is effective and fine.   Price 25 cts.

Italia. Of a bright orange-scarlet with a very broad golden yellow border; the flowers, which are produced on massive stems about sixteen inches long, are set well above the foliage; flowers of immense size, frequently measuring six inches across.   Price 20 cts.

Austria. A counterpart of "Italia" in every way, except in the color of its flowers, which are of a pure canary color, with but a few traces of small reddish brown spots in the center of the two inside petals.
                                                  Price 20 cts.

Alsace. A vigorous grower, with clear green foliage, four feet high; early and continuous bloomer. Spikes of flowers full, upon opening, are of a delicate sulphur yellow, changing very soon, however, to a creamy white without spots.   Price 15 cts.

Marigold. A rich orange color, with less scarlet or salmon than any so-called orange; flowers very large, of excellent form, petals one and three-fourths inches wide, foliage dense, glaucous green, seldom growing more than two and one-half to three feet high.   Price 15 cts.

Philadelphia. Two to three feet high, bright scarlet flowers, shading to crimson, five and one-half to six inches across, petals one and three-fourths to one and seven-eighths wide, not reflexed; a glowing color, pronounced by Mr. Wm. Falconer, as the finest red Canna in his collection, which includes all the best varieties known. It will supercede all others of similar shade.   Price 25 cts.

Queen Charlotte. A wonderful and decided novelty from Germany. Its main advance being its color, which is bright red; each petal bordered with a very wide band of bright yellow. Extra showy for summer use, and also a splendid winter bloomer.   Price 20 cts.

Nellie Bowden. A dwarf, large-flowering, pure bright yellow Canna. It is not spotted or marked with any other color, except that it has a touch of orange in the throat. One of the best attributes of this Canna is its free-flowering quality. It produces immense trusses of bright yellow flowers throughout the entire season.   Price 15 cts.

Charles Henderson. A magnificent variety. The flowers are deep, rich crimson, borne in compact, symmetrical heads, above a rich green foliage. The habit of the plant is dwarf—about three and one-half feet high—but the foliage is broad and massive. Taken all in all, this variety may well be considered the finest of recent importations.   Price 15 cts.

Florence Vaughan. The flower petals differ from most, even of the newest kinds, in their remarkable width and roundness, characteristics of the highest type to be sought for in these new kinds. In habit the plant has very broad, light green leaves, making a splendid typical effect. The flower spike is large; each flower opens broadly, and the color is a golden yellow, dotted with brilliant carmine.
                                                  Price 15 cts.

Madame Crozy. The flowers are of large size, of a dazzling crimson-scarlet, bordered with golden yellow. The plant is of vigorous growth, yet dwarf in habit. The foliage is of rich, cheerful green; flowers are produced in large branching stems, each stem being really a bouquet in itself.
                                                  Price 15 cts.

In May and June we will have of many of above sorts, nice plants, ten to twenty inches high, well grown in pots. These when planted out will bloom much sooner than dormant roots, and for immediate effect are very desirable. They cannot be sent by mail, but must go by express. Price, our selection in finest assortment, 25 cts. each, $2.00 per dozen. Buyer's selection, price on application.

RUDBECKIA.

Golden Glow (new double). A glorious new plant, and one that will find a place in every garden here and abroad; of fine habit, vigorous growth, with early, continuous and immense blooming qualities; it will be found excellent also for cut flower purposes, inasmuch as the stems are long, and often carry a dozen flowers furnished with beautiful foliage. A large vase filled with these flowers is a beautiful sight. One of the most distinct and beautiful, large growing, hardy perennials that have been introduced for years. In our grounds the past season it proved a great attraction, and we recommend it to our customers as a plant of great value. Hardy, of strong growth, long time in bloom, with flowers like large golden balls. We pronounce it one of the most satisfactory plants we have ever helped to introduce. Price, nice plants, 15 cts.; extra strong, 30 cts.

JUSTICIA.

Velutina. This new dwarf Justicia begins to bloom when the plant has only three or four leaves, and it is never out of flower afterward. If pinched back occasionally it makes a very dwarf, stocky plant, frequently covered with twenty to fifty large pink flower heads, lasting a long time. The foliage is also more persistent and highly ornamental, being heavy in texture and very velvety.   Price 15 cts.

VIOLETS.

Swanley White. A great favorite. Grown in pots or boxes, it is a beautiful sight; vigorous and healthy, bearing in profusion pure snowy white, perfectly double flowers, exquisitely fragrant. Lift and place in pots during September, and it will bloom elegantly during winter and spring. (See cut.)

Price 10 cts.; 3 for 25 cts.; 12 for 75 cts.

Lady Hume Campbell. This is a distinct and lovely new Violet that has become the most popular double variety among florists, thousands of them now being grown for cut flowers. Among the doubles it is beyond question the best for amateur culture, for the reason that it excels all others in strong, healthy growth, and it will produce under ordinary conditions, a wonderful abundance of flowers. Foliage large, clean and bright. The flowers are perfectly double, with most delightful violet fragrance.

Price 10 cts.; 3 for 25 cts.; 12 for 75 cts.

California (single). Its clearly outlined petals are the embodiment of graceful beauty. In color it is decidedly rich—a deep pure violet blue that does not fade but remains pure to the end. The flowers are exquisitely fragrant and borne on long stems. Price 10 cts.; 3 for 25 cts.; 12 for 75 cts.

Schoenbrunn. This single Violet is grown by the million for the Philadelphia market. Very desirable for the amateur. In mild climate, it will bloom all winter in open ground. Grown in open air at North, it will commence in September to push out its beautiful, dark fragrant flowers and will bloom till Thanksgiving in the garden, then with low frame and sash it will flower till Christmas, if covered with little extra covering on the coldest nights. Do not despair of raising Violets until you have tried this one.   Price 2 for 15 cts.; 5 for 25 cts.; 12 for 50 cts.

Marie Louise. The great market Violet. No other variety superior to it in color, which is of a very deep shade; blooms large and fragrant and produced in abundance. Price 10 cts.; 3 for 25 cts.

Princess of Wales. A grand new single sort—intensely fragrant. Little lighter than "California;" very desirable. Price 10 cts.; 3 for 25 cts.

Russian. A hardy sort for open ground culture at the North. Nearly all varieties of the fine double sorts are not hardy enough to stand northern winters without glass protection. This Violet has been grown for years, planted in the open ground;with no protection beyond covering the beds in the late fall with forest tree leaves. Flowers large, double, deliciously fragrant, and of a very deep shade of blue —one or two petals being dotted with red. Price 10 cts.; 3 for 25 cts.; 12 for 75 cts.

NOTE.—The Russian Violet is perfectly hardy. Other varieties require in northern states, good protection in frames if wintered for spring blooming. For house culture, pot them in September, and keep them in open air well watered, till freezing nights come, then place them in windows of a room where they will not have artificial heat, except perhaps the coldest nights, for while little frost will not injure, a hard freeze might do so; also, give air when ever it is above freezing out doors. Violets can be readily grown if above treatment is given them, but it is impossible to grow them in a house temperature of sixty-five to seventy-five, when they want only thirty-five to forty-five, with perhaps fifteen degrees warmer when the sun is shining on them. Try them with the cool treatment, not forgetting the air when warm enough.

HYDRANGEA.

Monstrosa (Improved "Otaksa"). One of the most valuable new plants that has been introduced in years. The trusses of flowers are of enormous size, even on the very small plants. Color, intense deep rose. Plant a strong free grower and remarkably free in bloom. With its clear solid pink color and immense blooms, this variety easily stands as the grandest of all Hydrangeas. (See cut.) Price, small plants, 20 cts.

Thomas Hogg. Another Japan production. A pure white variety of the old garden "Hydrangea," being identical with it except in color, which is of purest white. Hardy at North if protected with boughs; very fine for pot culture. Price 15 cts.; large, 60 cts. and $1.00.

Red Branched. This is one of the finest varieties in cultivation, with dark red branches that brighten as they near the flower trusses. The plant is of robust habit, and produces freely immense heads of deep rose-colored flowers. Price 15 cts.; large, $1.00.

Otaksa. A magnificent flowering shrub, with immense trusses of flowers, frequently twelve inches across; color, bright rosy pink, contrasting finely with other sorts. The plant is of low, bushy growth, and should have a covering of straw, leaves or evergreen branches during winter in northern climates. The low, sturdy growth, together with its immense blooms and beautiful foliage, make it the most desirable of all to grow in pots, tubs or vases for summer decoration. When so grown it can be wintered in cellar. Price 15 cts.; large, 60 cts. and $1.00.

Paniculata Grandiflora. One of the finest hardy shrubs in cultivation; the flowers are formed in large white panicles or trusses nine inches in length. The shrub grows to a height and breadth of three or four feet, and as the flowers slightly droop, few plants have the grace and beauty presented by this magnificent shrub. Perfectly hardy in all parts of the country; needs no protection of any kind. It should have all weak wood cut away each fall, and be well manured each season. (See cut.) Price, extra strong, 1-year field grown plants, 20 cts.; extra strong, 2-year field grown plants, 50 cts.; extra strong, 3-year field grown plants, 75 cts.; 4-year, tree shape, extra fine, $1.50; 4-year, bush shape, $1.25. All of above will bloom first year. The 3-year plants are extra fine, well branched, two to three feet high; all have extra good roots. The two, three and 4-year plants should go by express.

NOTE.—" Paniculata" is perfectly hardy in open ground. The other sorts, in northern states, are best grown in pots and tubs; they can then be wintered in cellars.

CHRYSANTHEMUMS.

We call special attention to our list of Chrysanthemums; nearly all are prize winners at the large shows in different parts of the country. If not acquainted with the improved new sorts of today, as compared with the Chrysanthemums of ten years ago, you will be astonished by a trial of a few of the best.

Lawn Tennis. Japanese reflexed, color of deep mauve, which is very attractive; build of blooms most perfect; of Japanese reflexed type; stems stiffest possible; foliage large and right up to bloom.
                     Price 15 cts.

Casco. An incurved Japanese. In color it is unique, being a light shade of garnet, exceedingly bright and attractive, causing it to hold a place not filled by any other variety of its season. A good, clean grower, unexcelled as a pot or bush plant.   Price 15 cts.

Mme. Ferlat. Pure white; petals regularly incurving; petals at times pointed gold. This is a fine new sort from France.   Price 20 cts.

Rachais. Bright bronze and yellow; enormous incurved form; one of the finest varieties of the year. Very full and substantial. The best plumed variety in the dark shades.   Price 15 cts.

The Harriott. Deep rose, silvery reverse, evenly arranged petals, making a well balanced flower; bloom of large size and great substance. This variety as a late pink has great value, being at its best from December 15th to 20th; strong stem and foliage. A fine Christmas flower.   Price 20 cts.

Little Chris (new). Small to medium flowers, pure white; a new break in Chrysanthemums, as it might be called a perpetual bloomer. Flowers at times in August and September, long before other sorts; then in winter and at Easter, flowers can often be cut. We offer it as a novelty of merit. (See cut.)

Price 35 cts.

Autumn Glory (new). A grand commercial pink, and unsurpassed as an exhibition variety. There is nothing like it in color, which is a deep reddish salmon upon opening, and gradually changes to a soft shrimp pink; of the largest size, and takes on great depth. A very easy doer; has no troublesome peculiarities.

Price 20 cts.

Miss Gladys Vanderbilt (new). Pearl white, with lemon-shaped center; an exceedingly fine thing, and a decided novelty both in form and

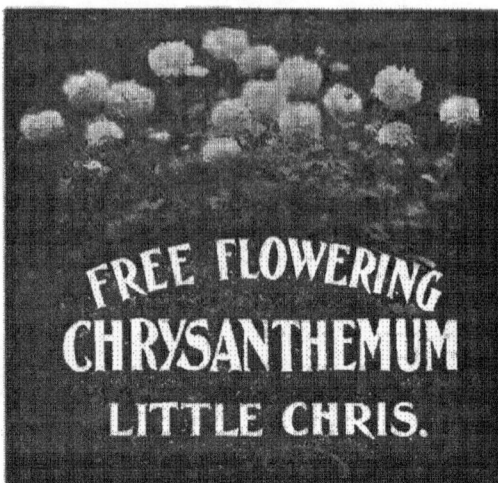

general build, and an extra good commercial or exhibition flower. Price 25 cts.

Yanoma. Innumerable inquiries come every year for the best very late white. For the present, "Yanoma" must be accorded this place; beside the fact that the flower is very beautiful and substantial, it is very easy to bring to perfection, which is a great point in its favor, as many of the late varieties are poor, defective growers; the form is rather flat, what we call the pure rosette form. Price 15 cts.

Elise Walker. Golden brown fimbriated; beautiful reflexed form; like a bright French Marigold. Pompon. Price 10 cts.

Mlle. Elsie Dordan. Soft pink, almost white; a perfect globe of recurved petals; extra fine both in color and form. Pompon. Price 10 cts.

Black Douglas. Dark crimson scarlet; extra fine; color gorgeous. Pompon. Price 15 cts.

Gold Lode. New early yellow color, very pure and bright. Excellent for cutting and a fine pot plant. Price 10 cts.

Miss Minnie Wannamaker. One of the best white varieties; flowers large, of great substance. Incurved at times, resembling a white ball. Price 10 cts.

Canary Bird (Pompon or Button). Flowers light yellow, and like small balls; blooms early. Very profuse. Price 10 cts.

Marion Henderson. A superb, large flowering, early sort. Pure gold in color. It is full of petals to the center, and lasts a long time; in fact, although very early, it can yet be made late, because it remains perfect on the plant for such a length of time. Price 10 cts.

Queen. This sort has taken more prizes than any other white sort; truly grand, when disbudded to one flower to stem. Price 10 cts.

Vivand-Morel. This grand variety has literally leaped into popularity the past year, and has done much to restore confidence in Continental varieties; color, tender rose; of very silky texture; petals long and straight, a fine grower; flower of large size and fine color, extremely beautiful. Price 10 cts.

Tuxedo. Beautiful shade of blended old gold and amber, the most novel in color of any of the varieties; a free bloomer. Price 10 cts.

Eda Prass. A fine, bold, recurving flower, of great substance and depth. When opening, of a delicate salmon, then changing to a creamy white, delicately shaded blush. One of the best for pot culture, dwarf and bushy. Price 10 cts.

Ivory. Snowy white, of perfect incurved form, very large; this is undoubtedly one of the most useful introduction of late years, as it is an early variety, of dwarf habit and free-flowering. Price 10 cts.

Golden Rod. Beautiful golden yellow flowers of fine shape, and borne in the greatest profusion, yielding more flowers than any yellow we ever grew. Price 10 cts.

J. Shrimpton. Bright crimson, color similar to "Cullingfordi," florets strong and stiff, flower of largest size and most perfect form. Price 10 cts.

Dorothy Toler (new). This is a gem, and while of only medium size, its great depth, perfect fullness, fine pink color and lasting qualities will make it rank as the leader among late pinks. Pure "Dahlia" form, gradually becoming reflexed. Price 10 cts.

Mrs. Henry Robinson. In the entire Chrysanthemum family there is no grander white variety, no matter when it blooms; it is of the most popular style—immense incurving broad channeled petals of great substance and the purest white. Price 15 cts.

Glory of Pacific. New pink sort, but entirely distinct from above. It is of magnificent size, with broad petals, which finally reflex; color changing to white, with rosy tints; grand grower. Price 15 cts.

E. G. Hill. Immense bloom, of brightest golden yellow; full and very double. Lower petals sometimes deeply shaded bright carmine. An elegant variety of strong habit. Price 10 cts.

ROSES.

HARDY GARDEN ROSES.

The following Roses are hardy, but it is better to protect in winter in northern states.

Evergreen Gem (new). A wonderful new hardy Rose, with sweetbriar fragrant foliage of a bronzy color; flowers on single stems, perfectly double; color, yellow, shaded buff, changing to white. This is a very strong climbing rose; will grow where any sort will, and is very fine for trailing on ground, for trellises, and fine for cut flowers. Price 60 cts.

South Orange Perfection. This is a gem, growing freely, and having multitudes of the most perfectly formed double flowers, about one and a half inches in diameter, soft blush pink at the tips, changing to white. Price 15 cts.

Pink Roamer. The single flowers, which are produced in close heads, are nearly two inches in diameter, bright rich pink, with almost a white center, which lightens up the orange-red stamens, producing an effect which, combined with fragrance, makes it one of the most valuable roses. Price 15 cts.

Manda's Triumph. This grand Rose is of free growth, luxuriant foliage, has large clusters of double pure white flowers, beautifully imbricated, two inches in diameter; sweetly scented. Price 15 cts.

Universal Favorite. Very vigorous. The double flowers are over two inches in diameter, and of a beautiful rose color, similar to the "Bridesmaid," and deliciously fragrant. Price 15 cts.

The habits of the above five new Roses are the same as the Rose Wichuriana, or Memorial Rose, of which they are hybrids. The growth is creeping, and therefore can be used to cover ground, stems of trees, pillars, posts, trellises or any other purpose desired, including pot culture. They are the hardiest Roses known, and will stand any climate or exposure, and will thrive in the poorest soil. The flowers are produced in the greatest profusion; deliciously fragrant and last a long time in perfection.

Wichuriana (New Japanese Creeping Rose, also known as Memorial Rose), A low, trailing species, its stems creeping on the earth as closely as Ivy, and forming a dense mat of very dark green lustrous foliage. The flowers are produced in greatest profusion, in clusters, after the June Roses are past, and continue during the season. Pure white, from one and one-half to two inches across, and very fragrant. Valuable for covering banks, rockeries, slopes and beds among shrubs; also fine for trellises, covering buildings, fences and stumps; extremely hardy. Price 10 cts.

Yellow Rambler (Aglaia). This new hybrid Noisette produces its flowers in the same manner as the popular "Crimson Rambler"—that is, in large pyramidal shaped trusses, frequently being from fifty to one hundred flowers in a truss. In color, it is a decided yellow, with moderate sized cup-shaped, nearly full flowers, which are sweetly scented. Its habit of growth is very strong, frequently making shoots of eight to ten feet in one season. This is as hardy as "Crimson Rambler," and possibly more so; not yet fully tested by us. Price 15 cts.

Crimson Rambler. This is undoubtedly the greatest acquisition in a climbing Rose introduced for many years. It is of vigorous growth, making shoots from four to ten feet in height during a season, and is consequently a most desirable climbing variety. As a pot plant it is unequaled for decorative purposes. The flowers with which the plant is covered in the spring are produced in large trusses of pyramidal form, and of the brightest crimson color. The blooms remain on the plant for a great length of time without losing their brightness. It is hardy, having withstood the test in exposed situations two winters, but it should have protection in extreme North. Price 15 cts.

Mme. Chas. Wood. One of the best Roses for general planting ever introduced; the flower is extra large, full and double; color, deep rosy crimson; sometimes brilliant scarlet with maroon shading. A constant and profuse bloomer. Price 20 cts.

American Beauty. The largest Rose and the most vigorous growing by far among continuously blooming Roses, throwing up very strong stems from the base of the plant, every shoot terminating with a single flower, which is of the largest size. Color, deep rose. In foliage and size of flower it resembles the large Hybrid Perpetual. It is as fragrant as the most fragrant of hardy Roses. Price 20 cts.

Magna Charta. A splendid sort. Bright clear pink, flushed with violet-crimson; very sweet; flowers extra large; fine form; very double and full; a free bloomer. Price 20 cts.

La France. One of the most beautiful of all Roses, and is unequalled by any in its delicious fragrance. Very large, very double, and superbly formed. It is difficult to convey any idea of its beautiful coloring, but the prevailing color is light silvery rose, shaded with silver-peach and often with pink. This variety is hardy for out-door culture, if slightly protected with evergreen boughs in winter. Price 20 cts.

Gen. Jacqueminot. Now known everywhere. Color, rich crimson; of fine shape and exquisite fragrance. Being very hardy, this variety is not only popular for winter forcing but is one of the best for out-door planting. This model and grand old variety holds its own against all new comers, and is undoubtedly the finest hardy Rose of its color. Should be in every collection. Price 20 cts.

Mme. Plantier. A hardy, pure white double Rose of the "Hybrid China;" one of the best white Roses there is. It is excellent for cemetery decoration, etc.; growth free and vigorous. Price 20 cts.

Coquette des Blanches. White; large and full; fragrant. Fine for cemeteries. Price 20 cts.

Mrs. Degraw. A superb Rose, and we can confidently recommend it as being hardy, equalling the the tender sorts in profusion of bloom. From early summer until frost it may be depended upon to produce flowers in abundance. In color it is rich, glossy pink, delightfully fragrant, and is such a strong grower that it is almost impregnable against attacks of the insects which usually destroy some of our finest Roses; thus it is certain to be unequalled for the garden and cemetery. Price 20 cts.

Baltimore Belle (climber). Blush, tinged rose and white; the flowers are very double, and are produced in great clusters. Whole plant is one complete mass of bloom when at its best. Price 20 cts.

Prairie Queen (climber). Dark rosy red, changing to pink. Price 20 cts.

Seven Sisters. Flowers in large clusters; varies in color from white to crimson. Price 15 cts.

Countess de Murnais (moss). Large, pure white, beautifully mossed; a splendid out-door Rose. Price 25 cts.

Henri Martin (moss). This is a magnificent Rose. The flowers are extra large, very double. The color is a deep, rosy carmine. Price 20 cts.

NOTICE.—We can furnish in April and May the following varieties from above list in strong two-year bushes: *Crimson Rambler, Yellow Rambler, Mrs. Chas. Wood, American Beauty, La France, Jacqueminot, Magna Charta, Mme. Plantier, Coquette des Blanches, B. Belle, Queen of Prairie, Seven Sisters, Persian Yellow* (a bright yellow variety). These thirteen varieties are all two-year hardy field-grown bushes, fine and strong, that will bloom finely if planted early. Price 50 cts. each; if by mail, 60 cts. each; but we advise express when possible.

TEA AND OTHER MONTHLY ROSES.

The following Roses are not hardy in open ground in northern states, but are fine for summer bedding or for the winter window garden.

Souv. de Jeanne Cabaud (new). A beautiful "Frenchy" Rose of exquisite color; looks like a bright pink Rose set in the heart of a large orange Rose; both colors are very bright. It is very large, of good substance and very full; a great novelty. Price 20 cts.

Mosella. While not by any means a yellow Rose, the general effect of the blending of the colors in this is such as to make it appear as a yellow Rose at a short distance. It is probably the finest Polyantha Rose yet introduced, and is a fitting companion to that very popular and useful variety,"Clothilde Soupert," but to which it is much superior in the shape of the buds and profusion of bloom. The habit of the growth is the same, while the flowers are white at the edge and chrome-yellow in the center. With protection, it is hardy, having withstood in the open ground a temperature of ten degrees below zero. Price 15 cts.

White Marshal Neil. Exactly like its parent, "M. Neil," except color, which is white, tinted lemon. A novelty recently received from Belgium. Price 25 cts.

Clothilde Soupert (Polyantha Rose). Flowers are borne in sprays. They are large, very double, and handsomely formed, round at first, but flattening as they expand; the outer petals are pearl-white, shading to a center of rosy pink, but varying sometimes on the same plant from pure white to deep silvery rose. It is fragrant and a constant bloomer. Grown as a pot plant, its free flowering habit and handsome form are sure to popularize it, while as a garden bedding variety it makes a compact mass of handsome foliage, with a multitude of buds and blossoms in varying shades of delicate white and rose. Hardy in open ground, with protection. Price 10 cts.

Maman Cochet. This is one of the most beautiful new Tea Roses that has been introduced in years. The growth is vigorous,with rich healthy foliage and extra large flowers on long stout stems, very double and simply exquisite when in bud or half blown. The color is a deep rosy pink, the inner side of petals silvery rose, makes charming bunches of long-stemmed flowers when cut. Fine for either pot culture or out-door planting. This Rose is so much better than "Mermet" for bedding out, or for house culture that we drop that sort from list. Price 15 cts.

Hutton. The stock of this fine Rose came to us through an amateur. We cannot give its history, but have never before seen it, nor can we find it described in any American catalogue. Its habits are similar to Polyantha Roses, and of a bright crimson color; entirely distinct from any Rose we have seen; should be in every collection, as it is sure to become a favorite. Price 10 cts.

Marshal Neil. No yellow Rose yet produced equals it. The flowers are of the largest size, highly fragrant, and of the richest and deepest golden yellow; the flowers remain in perfection several days. It is a vigorous grower.   Price 15 cts.

Hermosa. A grand Rosé and one which no one can afford to be without. The flower is cupped, finely formed, always in bloom. Soft shade of pink, fragrant, but not a Tea.   Price 10 cts.

Perles des Jardins. Flowers very large, full and well formed; color, beautiful straw or rich canary; vigorous growth, extra fine.   Price 15 cts.

Duchess de Brabant. One of our most valuable summer and winter blooming varieties, equally good for either season, free bloomer; color, light carmine, tinged with violet; buds full and very fragrant. Extra for pots or bedding.   Price 10 cts.

Papa Goutier. A lovely Tea Rose, which has proved to be all that has been claimed. The blooms are large and long, with thick, broad petals of a dark carmine-crimson. The inner petals are a bright rosy carmine, and light up well at night. It is delightfully fragrant and combines all the qualities of the " Bon Silene." Exceedingly free-flowering, summer and winter.   Price 10 cts.

Marie Guillot. White, tinged with delicate shade of lemon; large, full and beautifully imbricated in form; one of the finest white Teas; the perfection of form in Tea Roses; highly fragrant. This Rose is so much better than the " Bride" that we drop that sort from our list.   Price 10 cts.

Couquet de Lyon, the Yellow Hermosa. No Rose like it for freedom of bloom. The plant is a fine grower, and carries numbers of flowers on every shoot. A fine yellow Rose, large, not at all formal, but very attractive. Called " Yellow Hermosa" from its free-flowering character. Price 10 cts.

Marion Dingee. This promises to be of extraordinary value. It is a strong, vigorous grower, making a handsome, graceful bush, with large, thick, deep green leaves. The flowers are large, beautifully cup-shaped, moderately full, and borne in wonderful profusion all through the growing season. The color is perhaps its most remarkable feature. It is a deep, brilliant crimson, one of the darkest, if not the very darkest and richest colored Tea Rose in existence.   Price 15 cts.

DAHLIA.

Gilt Edge. Won first prize as the best new Dahlia. The petals are recurved like a Chrysanthemum; color ivory-white, deeply edged with rich golden yellow; entirely unique and most beautiful. Price 15 cts.

Grand Duke Alexia (new). Without exception, the most unique Dahlia, and the grandest ever introduced. Frequently eighteen inches in circumference, most beautiful in form; none finer; color, ivory-white with pink center; the petals, which are tubular in form, being just sufficiently opened at their extremity to show a faint shade of blush or peach color, greatly enhancing the whole effect. A most admirable flower for cutting, as it remains fresh much longer than any Dahlia we know of. The plant is an elegant and robust grower, with large, dark foliage.   Price 25 cts.

Clifford W. Bruton. This superb new variety is acknowledged by all who have seen it to be the finest yellow decorative Cactus Dahlia ever produced. It is a very strong, sturdy grower, and an extremely profuse bloomer for so large a flower. The flowers are large, five to six inches in diameter, and perfectly full to the center. The color is a solid pure yellow of the clearest and richest shade. On this account the flowers can be cut at any stage of development, and as they are borne on very long graceful stems, they are invaluable for cutting or exhibition purposes.   Price 25 cts.

Iridescent. A beautiful Cactus variety with a combination of colors that is difficult to describe, the ground color being a clear orange overlaid with red and suffused with pink, giving a soft yet brilliant red effect with a blue reflex. The odd and beautiful coloring of this sort has caused the greatest praise everywhere when exhibited. Very rare.   Price 30 cts.

Wm. Agnew. The grandest red Cactus Dahlia ever produced. The flowers are of large size, sometimes measuring more than seven inches in diameter, and always full to the center; of perfect form and of exquisite quality and finish, while the color is the richest shade of intense dazzling red. The petals are very long, the outer rows being beautifully twisted; the plant is a strong, symmetrical grower, and, owing to perfect habit, will be especially valuable for every purpose.   Price 30 cts.

Nymphæa. By far the most delicately beautiful Dahlia ever introduced, and is more extensively grown for cut flowers than all others combined. The flowers are medium large size, always full to the center, and so clearly resembles the ideal pink Water Lily as to suggest the name. The color is a clear, distinct, light shrimp pink, tinted lighter toward the center. The plant is a strong grower, of medium height, shrubby growth, and is the most profuse bloomer of any variety, being a mass of bloom from June until frost.   Price 30 cts.

Vivid. A very showy pompon giving a profusion of perfectly formed flowers of the most intense scarlet.   .   Price 15 cts.

Dahlias in variety. We have many varieties not listed above. We will send postpaid, two varieties in distinct colors not named, for 25 cts., or 5 for 50 cts.

The recent improvement in this class of plants is causing a great demand for them. No finer ones in the trade than the first five on list; these are all of the newer type and every one a prize winner.

ASTERS.

Comet (White Giant). Elegant.
Price 50 ẻs. per doz.
Comet. Pink, medium height.
Price 50 ẻs. per doz.
Comet. Purple, medium height.
Price 50 ẻs. per doz.
Semple's White. Pure white.
Price 50 ẻs. per doz.
Mary Semple. Pale pink, extra.
Price 50 ẻs, per doz.
Lavender Queen. Semple's in pale lavender.
Price 50 ẻs. per doz.
Semple's Mixed. Assorted in above kinds.
Price 50 ẻs. per doz.
Fancy, in many colors and kinds,
Price 50 ẻs. per doz.
Truffaut's Perfection. White, a grand sort.
Price 40 ẻs. per doz.
Truffaut's Perfection. Bright pink.
Price 40 ẻs. per doz.
Truffaut's Perfection. Purple.
Price 40 ẻs. per doz.

Truffaut's Perfection. Dark red.	Price 40 ẻs. per doz.
Truffaut's Perfection. Above four kinds mixed.	Price 40 ẻs. per doz.
Mignon. Pure white, medium size.	Price 50 ẻs. per doz.
Mignon. Finest sorts, mixed.	Price 50 ẻs. per doz.

NOTE.—Our Aster plants are all grown from choicest seed, and grown in pots, ready for shipment after April 15th. For a full description of each sort see Seed Department.

FARFUGIUM.

Grande. A low growing foliage plant, remarkable for its shiny dark green foliage, which is irregularly blotched with bright yellow, and sometimes with white and rose. Easily grown, and a native of China; not a new plant, but not as well known as it should be, as it is a splendid window plant. Give plenty of water when in growing condition. (See cut.)
Price 30 ẻs.; large specimens, 75 ẻs.

FREESIA.

Refracta Alba. Too much cannot be said of this beautiful bulb for forcing. Pure white, tube-shaped flowers, having a most delicious perfume. A beautiful house plant, and grows easily in any good, rich, sandy soil. Flowers will keep two weeks after opening. One of the few Christmas flowers, and as soon as better known will be one of the most popular holiday decorations; also, it will be grown in large quantities by florists for cut flowers. Bulbs should be treated the same as winter-flowering Oxalis, i. e., kept in dry earth or sand until August, then potted off, three to six bulbs in a six-inch pot; these will bloom in December or January. Later potted, will, of course, bloom later. Like Oxalis, the bulbs increase rapidly. Price, extra selected bulbs, 5 ẻs., 6 for 25 ẻs.; fine large bulbs, 12 for 20 ẻs.; small bulbs, 10 ẻs. per doz.

HELIOTROPE.

Czar. A vigorous grower, with dark stiff stems carrying enormous trusses of bloom of an intense deep violet with white eye.
Price 30 ẻs.
Czarina. Strong robust stems and extremely large trusses of extra-sized flowers of deep indigo blue with white eye.
Price 30 ẻs.

The above two new giant Heliotropes are selections from the new mammoth type. Very large florets and immense trusses, often ten or more inches across; grand novelties.

Mrs. Burgess. Color, fine dark violet; one of the best dark sorts, in fact one of the best of all Heliotropes. Good for pots; fine for bedding in open ground, and extra good for cutting; odor exquisite, and it produces flowers in abundance. In last twenty years we have had on trial many varieties of Heliotrope, but we pronounce this the best, either for pots, bedding or cut flowers. Price 10 cts.; larger, 20 cts.

White Lady. The nearest approach to pure white yet in the Heliotrope; truss very large, growth compact; a decided acquisition. A slight improvement on "Snow Wreath." Price 10 cts.; larger, 20 cts.

TENDER VINES AND BASKET PLANTS.

The following vines are for house culture or summer planting—not hardy.

Asparagus. See Page 23.

Maderia Vine. Too well known to need description.   Price 10 cts.; 4 for 25 cts.

Saxifraga Sarmentosa. The old-fashioned basket plant, better known as Strawberry Geranium. Ornamental foliage.   Price 10 cts.

Vinca (Variegata). Rapid grower; leaves glossy green, margined with white.   Price 10 cts.

Mesembryanthemum Wax Plant (or Rock Pink). Curious succulent plant, similar foliage and flower to above. Flowers pink; an old variety, but fine for baskets.   Price 10 cts.

Mesembryanthemum Grandiflora Alba. New, large flowering, pure white wax plant, of a trailing habit, and especially fine for hanging pots or baskets. The foliage is round and fleshy, and the blossoms very large, two to three inches across, and exceedingly beautiful.   Price 10 cts.

Lophospermum Scandens. A handsome climbing plant of rapid growth, with heart-shaped leaves and rose-colored, Gloxinia-like flowers. Its rapid growth and fine foliage, with its beautiful flowers, combine to make it one of the best plants for covering any unsightly object or for large vase on the lawn. For a climber on the veranda it has no equal. Does not go well by mail.   Price 10 cts.

Lotus Peiyorensis (Coral Gem). This charming plant seems to be unknown in this country, yet it is so well thought of abroad that a prominent horticultural journal color-plated it. *The Garden* says of it: "Its slender branching habit is most striking, and the silvery foliage even more so. Without its flowers it might almost be described as a Silvery Asparagus." The bright coral-red flowers, measuring two inches in length, are really beautiful; its drooping habit also makes it useful for hanging-baskets. Easily grown. (See cut.) Price 10 cts.

Hoya Carnosa (Wax Plant). Has thick, fleshy leaves growing moderately fast and bearing umbels of beautiful flesh-colored flowers.   Price, fine plants, 15 to 30 cts.

Myrsipyllum Asparagoides (Smilax). A climbing plant unsurpassed in the graceful beauty of its foliage; valuable for bouquets and decorations.   Price 10 cts.

Passion Vine, Constance Elliot. This beautiful variety is hardy with slight protection. A climbing plant rivalling some of the Clematis in size, color and profusion of bloom, and surpasses them in vigor of growth and delicious fragrance, and are pure ivory white.   Price 10 cts.

Ivy Silver Stripe (new). Fine foliage. Each leaf heavily bordered with pure white. The finest fancy Ivy.   Price 10 cts.

Cuphea Platycentra (Cigar Plant). The tube of the flower is scarlet, with the end part white and crimson, having somewhat the appearance of a miniature lighted cigar; not a vine, but a trailing or basket plant.   Price 10 cts.

Passion Vine, John Spalding. A beautiful variety, foliage prettily marked with bright golden yellow. Flowers same as "Constance Elliott."   Price 15 cts.

English Ivy. Too well known to need description.   Price 10 cts.; large, 30 cts.

Honeysuckle, Golden Leaved. Marvelously effective. It bears innumerable sprays of the most perfect little leaves that are covered with a net work of gold and green, often marked with pink also. The gold deepens at the tip of the sprays, and there is nothing finer to combine with bouquets; also quite hardy in open ground.   Price 15 cts.

Lantana, Mrs. McKinley (New Weeping). For vases, baskets or pots, for which it is particularly recommended. It is a neat, handsome grower, producing a great abundance of leafy vines and rosy pink flowers and always attracting a great deal of attention. The flowers are borne in pretty clusters, and it blooms quickly and abundantly in small pots. A charming plant for the window garden; grows nicely in boxes, pots or baskets, and is scarcely ever without flowers.   Price 15 cts.

Glechoma Hederacea var. (Nepeta). A variegated form of the common ground Ivy, which is prettily marked with pure white on a bright green ground. It is hardy and of vigorous growth like the parent variety, but unlike it, the leaves of this form have a refreshing fragrance similar to Mint. It will no doubt become one of our most useful plants for festooning work, such as drooping from hanging baskets, vases, fancy pots, etc., etc.; its graceful, rapid growth rendering it unequalled for this purpose, while we have no doubt it will also be largely used in cemeteries to cover graves, as it is entirely hardy and we think more ornamental than almost anything in use for that purpose at present. (See cut.)

Price 10 cts. each; 3 for 25 cts.

Jessamine Grandiflorum (Catalonian Jessamine). A valuable winter flowering plant, blooming without intermission from October to May. The flowers are pure white, most deliciously fragrant; used extensively by all bouquet makers. Price 20 cts.

Ivy German (or Parlor). A well known rapid climber; for pot culture or vases. Price 10 cts.

Cobea Scandens. A well known climbing plant of very rapid growth and having large blue bell-shaped flowers. Used for covering trellises and unsightly places, for which its rapid growth makes it very valuable. Price 15 to 30 cts.

Manettia Vine. In the house it can be trained all around a window, and will bloom both summer and winter. In the garden its charming beauty surpasses everything. Flowers intense scarlet, tipped with yellow, each flower keeping perfect over a month before fading. It can be trained on a trellis, strings, or used for drooping from hanging baskets; in any way a perfect mass of flowers and foliage from the root to the tips of the branches. Price 10 cts.

GLADIOLI.

The Gladiolus is the most beautiful of the summer bulbs, with tall spikes of flowers, some two feet or more in height, often several from the same bulb. The flowers are almost every desirable color. As cut flowers, they are the most lasting of anything we know. By cutting the spikes when two or three of the lower flowers are open, and placing them in water, the entire spike will open in the most beautiful manner. Set the bulbs from six to nine inches apart, about three inches deep. In the fall, before hard frost, take up the bulbs, remove the tops, leave to dry in the air a few days, and store in some cool place, secure from the frost until spring. Our mixed varieties are unusually fine.

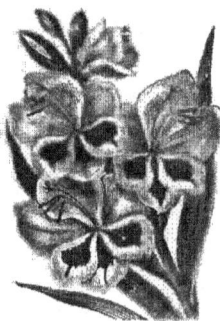

Lemoine's Hybrid. For vivid and rich orchid-like coloring, this beautiful, half-hardy class of Gladioli has no equal. Our bulbs were raised from selected seed, saved from the finest named varieties, selected for their distinct and brilliant colored flowers. Many of these are equal to the finest named sorts. (See cut.) Price, extra fine mixed, 5 cts., 50 cts. per doz.

Hesperide. Profusely blotched, and flaked bright rosy salmon on a pure white ground; a finely colored sort. Price 15 cts.

John Bull. Ivory-white, tinted with yellow and lilac.
Price 10 cts.

Le Poussin. Light red, with large white throat; a beautiful variety. Price 10 cts.

Isaac Buchanan. The best yellow variety grown; a great novelty. Price 10 cts.

Snow White. The nearest pure white of any variety yet produced. Extra fine. Price 20 cts.

Lord Byron. Brilliant red, blotched white; a very showy sort. Price 10 cts.

Childsii. New, strong growing, large flowering varieties, mixed. A great improvement on Lemoine's. Price, in mixed colors, 10 cts.

Gladiolus (fine mixed). The bulbs are all fine varieties. We send them out without names. One of the finest lot of Gladiolus ever offered at this price. Price, 40 cts. per doz., by mail, prepaid.

ALYSSUM.

Mammoth. The finest of all large growing Alyssums. Small plants, planted in open ground will make immense clumps, for summer cut flowers. Also grand as a pot or vase plant; flowers white, semi double, of twice the size in truss and floret of any sort we have ever seen. Its long, strong stems makes it very fine for cut flowers. First-class for summer or winter use. Price 10 cts.; 4 for 25 cts.

Double Tom Thumb. This variety has the same dwarf habit as the single, but its trusses are much longer and the florets double. The double flowers hold on so long that the plant is always in bloom, never having that ragged and unattractive appearance of old varieties when out of flower. Price 10 cts.; 4 for 25 cts. Either sort, 60 cts. per doz.

MISCELLANEOUS.

Azalea. Elegant pot plants, much used for Easter decorations. Though attaining a good size, commence to bloom when very small. The flowers are large and very beautiful, always much admired, and range from white to deep crimson in color, with all intermediate shades and markings. Easily grown in any window. Our plants are fine ones; all in 5, 6 and 7-inch pots, well branched and full of buds; will bloom profusely through the spring months.　Price 60 cts. to $1.00; extra sizes, $1.25 to $1.50.

Ardisia, *Crennlata.* A very ornamental greenhouse plant, dark evergreen foliage producing brilliant red berries; a first-class house plant.　Price, quite small, 10 cts.; strong plants, 40 cts.

Agave, *Marginata.* Beautifully variegated green, yellow and white.

　　　　　　　　　　　　　　　Price 25 cts., 50 cts., $1.50 to $2.50 each.

Agave, *Americana* (Century Plant). Dark green; strong grower.

　　　　　　　　　　　Price, small, 25 cts.; large specimens, $1.00 to $3.00.

Agapanthus *Umbellatus, Lily of the Nile.* This is a splendid ornamental plant, bearing large clusters of blue flowers on long flower stalks, and lasting a long time in bloom. There is no finer plant than this for out-door decoration, planted in large pots or tubs on the lawn, terrace or piazza. It does well in the house or greenhouse in winter, requiring but slight protection.　Price 25 and 40 cts.

Amaryllis, *Belladonna.* A noble flower from the Cape of Good Hope; perfectly hardy south of Washington, and even further north if planted a foot deep and covered. The flowers appear in the fall, and are of the most beautiful satiny rose imaginable, and of exquisite fragrance. They are produced in umbels of six to twelve flowers in a cluster.　Price, extra strong bulbs, 50 cts.; small, 20 cts.

Aloysia Citriodora, *Lemon Verbena.* Indispensable for fragrance of its leaves. Price 10 cts.

Bouvardia, *Davidsonii.* White; often delicately tinted pink, profuse bloomer.　Price 15 cts.

Bouvardia, *Alfred Neuner.* The flowers are rather larger than those of the single flowering, and composed of three perfect rows of petals of the purest waxy white color, each floret resembling a miniature Tuberose.　Price 15 cts.

Calla, *Ethiopica.* The well known house plant. This Calla is the true large flowering sort; the grandest of all.　Price, small, 10 cts.; larger, 25 cts., 50 cts.

Calla, *Little Gem.* Grows from eight to fifteen inches high. A miniature Calla, with flowers from one-half to two-thirds the size of the old variety. It commences flowering when quite small. We had hoped great things for this sort, but find upon equal trial that it is in no way superior to the larger sort.

　　　　　　　　　　　　　　Price 10 cts.; larger, 25 cts.

Centaurea, *Gymnocarpa.* Attains a diameter of two feet, forming a graceful rounded bush of silver gray.　Price 10 cts.

Coleus. In twelve fine varieties.　Price 10 cts.; $1.00 per doz.

Cactus, *Nameless* (or Dr. Regal). Night-blooming, and without doubt the finest of the family. Flowers very large, sometimes measuring twenty inches in diameter; of the purest white. Originated in St. Petersburg. A handsome slender stemmed species, which ought to be grown much more generally than it is at present, for it is one of the finest of the night-flowering Cacti. The flowers are twelve to twenty inches in diameter, with creamery white lanceolate petals, with an outer fringe of narrow yellow sepals. The surface of the stem being covered with irregular tubercules, not distinctly ridged as in most other species.　Price 20 cts.

Cactus, *Epiphyllum* (or Lobster Cactus). A wonderfully free-bloomer, frequently blooming three or four times during the year. Of drooping habit. Flowers bright pink. Price, small plants, 15 cts.

Cactus, *Cereus Grandiflora* (Night-blooming Cereus). Flowers enormously large, and of rarest beauty. Pure white, and opening only at night.　Price 20 cts.

Cyclamen, *Persicum.* The Cyclamen is particularly adapted to window culture; it blooms abundantly and for a long time in the winter and spring. The flowers are borne on long stems above the foliage and are very showy. See colored plate, last page of cover.　Price, strong plants, 15 cts.

Cyclamen, *Persicum Giganteum.* Flowers of very large size; an improvement on older sort. An extra fine strain; two colors, red and white; fine plants.　Price, each, 30 cts.

Cyperus Alternifolius, *Umbrella Plant.* An ornamental grass, throwing up stems about two feet high, surmounted at the top by a whorl of leaves, diverging horizontally, giving it a very curious appearance. Splendid for the center of vases, or as a water plant; also a fine pot plant.   Price 15 cts.

Carex Japonica Variegata. The exquisite gracefulness of this decorative gem will find a place for it among Palms, Ferns and Dracænas. It is extremely easy to grow, will stand heated atmosphere with impunity, and can be carelessly handled without being damaged.   Price 20 cts.

Dracæna Indivisia, *Dragon Tree.* One of the most desirable of our ornamental foliage plants for decoration, either in or out doors. Its fine foliage renders it very useful for the window garden, planted as a center piece in a rustic stand, jardiniere, or window box, or for summer decoration in vases and ornamental beds; also especially well adapted for centers of vases, baskets, etc.; the bright green, narrow foliage contrasting well with other foliage and flowering plants.   Price 25 cts., 50 cts.

Daisy, *Giant White or Mammoth Paris Daisy.* A new large flowering, semi-double variety of the true "Marguerite" of the Paris markets; large, pure white flowers, three inches in diameter, with a double row of petals surrounding the golden yellow disk. It is a vast improvement, in size of flower, on the old "Marguerite," being twice as large and more numerous petals; lasting longer. It makes a good display in pots during the winter months.   Price 10 cts.

Daisy, *Hallerii.* New golden Marguerite, similar to above, except color, which is a beautiful light yellow, ever blooming.   Price 10 cts.

Daisy, *Snow Crest.* Double white English Daisy. A most distinct variety and an improvement on all existing sorts. Its habit of growth and size of flowers, in comparison with older varieties, is gigantic. The flowers,which are borne on stout, stiff stems from six to ten inches long, are of the purest white and full to the center. When fully developed they rise to a conical or sugar-loaf form, and well grown specimens will cover a silver dollar. It makes an admirable hardy garden plant or may be forced for cut flowers under the same treatment as Violets.   Price 10 cts.

Daisy, *Blue Celeste.* A daisy flower of intense blue, yellow center; blooms in winter.   Price 10 cts.

Ficus Elastica (India Rubber Tree). One of the best plants for table or parlor decoration; its thick, leathery leaves enable it to stand excessive heat and dryness, while its deep glossy green color always presents a cheerful aspect. The plants we offer are in fine order and of a size to be useful immediately.   Price 50 cts., $1.00 and $2.50 each.

Feverfew, *Little Gem.* The great value of the old Dwarf Feverfew for cut flowers during the summer and fall is well known to florists and others. This variety which we now offer surpasses it in every way. It is very dwarf, attaining a height of only twelve inches. The flowers are larger, of more perfect form and of the purest white.   Price 10 cts.; 3 for 25 cts.; 12 for 75 cts.

Forget-me-not, *Perfection.* A fine variety for winter bloom; deep blue and in flower from January to May.   Price 10 cts.

Genista Racemosa. A beautiful yellow flowering greenhouse shrub, flowering so freely as to literally cover the plant with its rich golden yellow flowers when well grown. It is grown in large quantities by the leading florists for decorative purposes at Easter time.   Price 10 cts.; larger, 25 and 50 cts.

Grevillea Robusta, *The Silk Oak.* A beautiful plant for decorative purposes; quick growth and of very easy culture. The leaves are a light bronze color, the tips being covered with a soft down resembling raw silk, from which it derives the name of "Silk Oak." For the house it is unsurpassed, as it needs little attention; flowers of a bright orange color.   Price 15 cts.

Hedychinm, *Coronarium* (Butterfly Lily). This magnificent tuberous-rooted plant flourishes on low, moist land, in half-shade, where it will form a dense clump three or four feet high. From August, onward, a mass of fragrant, pure white flowers, looking almost like large white butterflies, borne in large clusters, terminal on every stalk. Not hardy; should be taken up and treated same as Cannas in the winter. This plant is now being boomed by some florists under the name of "Myrosma Cannæfolia, or White Canna,"—and a new plant, but we have sold it under its true name for past five years.   Price 20 cts..

Hibiscus, *Aurauticum.* Large, double, orange-colored flowers; an early and profuse bloomer. One of the best.   Price 20 cts.

Hibiscus, *Grandiflora.* Rich, glossy foliage, blooming profusely during the summer, literally covering the plant with scarlet-crimson single flowers.   Price 20 cts.

Hibiscus, *Miniatus Semi-Plena.* Large, semi-double, brilliant and attractive; bright vermillion-scarlet; ought to be in every garden.   Price 20 cts.

Lantana, *Alba Perfecta.* Pure white, compact habit; fine.   Price 10 cts.

Lantana, *Michael Schmidt.* Flowers fresh and sparkling, of a brilliant yellow, passing into purple-vermillion.   Price 15 cts.

NOTE.—The Hibiscus and Lantanas are fine as house plants, but their greatest value is when bedded out. We recommend our customers to give them a trial. They can be wintered in cellar, same as an Oleander.

Lemon, *Sicilly.* Similar in growth to Japanese Orange, but fruit is of immense size, weighing from one to three pounds; fruits when quite small; a fine house plant.   Price 25 cts.

Lopesia Rosea. Fine red flowering plants, blooming continually from November to April; fine for cut flowers.   Price 10 cts.

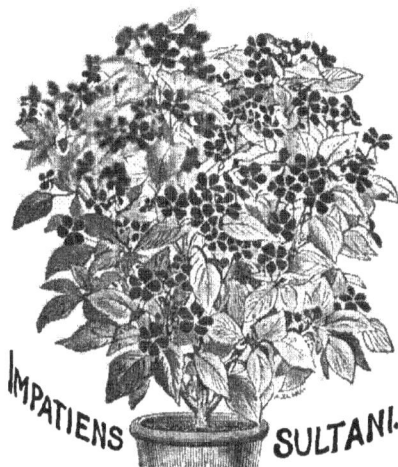

Impatiens, Sultani (sometimes called "Patience Plant"). This is one of the most distinct and beautiful of flowering plants of recent introduction, either as a winter blooming pot plant, or for the border in summer. It is of compact, neat habit of growth, with good constitution, and a perpetual bloomer. The flowers are of a brilliant rosy scarlet, peculiarly distinct, but most effective hue, about one and one-half inches in diameter, and are produced singly or twos and threes from the axis of the leaves, especially toward the summit of the stems, but so freely that a well grown specimen appears to be quite a ball of fire. We recommend this highly. (See cut.)   Price 10 cts.

Impatiens, Queen Corola. For a blooming plant that will flower the full 365 days in the year, and thrive well as a window plant, none equal the well known "Impatiens Sultani." This new sort is identical in growth and free blooming quality, but having flowers in color a beautiful salmon, suffused with rose, base of lower petals blotched white, stamens and pistils garnet and purple. The whole flower irregularly burnished, as in tinsel.   Price 10 cts.

Linum Trigynum. A native of East Indies. This beautiful showy old plant is well deserving of a place in every collection. It is often called a greenhouse plant, but under such conditions its beauties are never developed; it should be grown during summer in open border, taken up in the fall and re-potted. It is of dwarf, free growth, and the flowers are large, bright golden yellow, and disposed in large racemes. When well treated, they will produce their gay blossoms the entire winter.   Price 15 cts.

Orange, *Otaheite* (new from Japan). A new variety which is suitable for pot culture, and which will with anyone prove one of the most desirable pot plants it is possible to possess. It is not an uncommon thing to see little cutting plants five or six inches in height, full of bloom, and even bearing one or two medium-sized oranges. The fruit at their best are not more than half the size of an ordinary orange, but are very bright and beautiful in color and delicious in quality; but it is for its flowers more than its fruit which commends it to general cultivation. When it blooms it is so full that it seems to be all flowers. The pure waxy white blossoms emit a delicate, yet powerful fragrange, which is surpassed by no other flower.   Price, 1-year plants, 20 cts.; 2-year plants, 50 and 75 cts.

Orange, *Trifoliate.* The most hardy of the Orange family, and will stand in open ground, except in extreme North.   Price, 2-year pot plants, 25 cts.

Swainsonia, *Galegifolia Alba.* Pure white. Desirable ever-blooming plant with flowers produced in sprays of from twelve to twenty flowers each, the individual blooms resembling a Sweet Pea. As a decorative plant for the window or conservatory we know of nothing that will give as much satisfaction as the Swainsonia. Its easy culture, growing in almost any light position, its freedom of bloom, rarely ever being out of bloom, and the grace and beauty of the flower and entire plant will undoubtedly make this one of the most popular plants in cultivation. (See cut.)   Price 15 cts.

Stevia Nana. An improvement on "Stevia Compacta;" similar in every respect except in height, which is considerably less than the old variety, which makes it more desirable. In full bloom at Christmas.   Price 10 cts.

Strobilanthes, *Dyerianus.* The undulation on the surface of the foliage is furnished with a blueish metallic hue, shading into light rose with a light green margin. It surpasses the finest Coleus or Begonia in the exquisite coloring of the leaves, besides which the racemes of violet blossoms with which the plant is covered in mid-winter greatly enhances its beauty; a valuable bedding and decorative plant.   Price 10 cts.

Sansevelria, *Zealanica.* A beautiful plant, splendidly adapted for the decoration of drawing-rooms and halls, as it stands drought and dust with impunity. The leaves grow to a length of three to four feet, and are beautifully striped cross-wise with broad white variegations on a dark green ground. It is a rare and beautiful plant which should be abundantly grown for positions out of the reach of sunlight, where other plants will not thrive. When you consider that it may be placed in any position in any room and do well, its great usefulness is at once apparent. It has beauty for decorative purposes which other plants do not possess. (See cut.)

Price 10 cts.; larger, 25 cts.

Salvia Splendens, *Compacta.* Also known as Bedman, Bonfire, etc. A splendid new French variety, with very brilliant scarlet flowers and of comparatively dwarf and compact habit. Its blooms appear one month earlier than the old variety, "S. Splendens," and produces flowers in succession until frost sets in. An excellent plant for masses. Its early blooming qualities will be appreciated by all lovers of this fine plant; it also comes true from seed. We now grow this wholly in place of the old sort.   Price 10 cts.; three for 25 cts.; twelve for 75 cts.

Tuberose, *Excelsior Pearl.* A grower of cut flowers for the New York market, who has long been noted for his exceedingly fine florets of " Pearl " Tuberoses—which command a higher price than any other in the market—some fifteen years ago, when " Pearl " Tuberoses were first sold, selected the offsets from the bulb that had produced an exceptionally fine spike of large, beautifully formed flowers. The increase of this bulb has been the only kind grown at this florist's place, and he has guarded the stock with the utmost care, with the result that in the past fifteen years not one single bulb has ever differed from the original characteristics. In order to designate this special stock, we call it the " Excelsior Pearl." Price, fine blooming bulbs, 2 for 5 cts.; 25 cts. per doz.; extra strong, selected bulbs, 5 cts.; 50 cts. per doz.

HARDY PERENNIALS.

The growing interest in hardy plants has induced us to enlarge our list so as to include the most desirable varieties. We also classify under above heading, that buyers need have no difficulty in knowing what plants are hardy. The plants in following list are of easy culture, and once obtained will bloom each year with increasing beauty, and by a good selection you can have the most varied of nature's floral production during the entire season.

Astilbe, *Chinensis.* Very hardy and the finest Astilbe yet produced. The immense, yet graceful and airy sprays of a new delicate flesh-color are produced in immense profusion, literally covering the plants with flowers. The handsome dark green foliage adds effective color for cutting purposes, for which it will certainly be grown extensively.   Price 25 cts.

Achillea, *The Gem* (or Pearl). One of the most popular hardy plants. The blooming quality of this splendid novelty is something remarkable, flowering in great profusion the first summer, while the second season it will bloom three to five times as much; is entirely hardy, and will last for years. The flowers, which are borne upon erect footstalks, are of the finest white, and closely resemble a Pompon Chrysanthemum. A grand acquisition for cutting purposes. Succeeds anywhere, and flowers nearly the whole season; one of the most valuable snow white flowering plants introduced for years. Fine for borders; for cut flowers; also, unsurpassed for cemetery planting. (See cut.)

Price 10 cts.

Eulalia Gracillima Univittata. The most beautiful and useful of all the hardy grasses. Narrow green leaves with a silvery white mid-rib. This plant is of most graceful habit and is very beautiful for decorative purposes and the center of vases, as well as making an attractive lawn plant. One of the finest hardy plants in existence.   Price 15 cts.

Funkia, Day Lily. A handsome, showy plant, with beautiful large broad leaves. Flowers large, pure waxy white, borne in large trusses. Very fragrant.   Price 20 cts.

Campanula Persicæfolia. One of our finest hardy plants. Flowers in June, sending up fine spikes of pure white bell-shaped flowers. Their purity has also given this flower the title of "Angel Bells." Very hardy; roots should be divided once in two years. Grand flowering hardy plant. (See cut.)   Price 20 cts.

Monarda, Didyma, (also known as Bee Balm, Oswego Tea, Horse Mint, etc.) A sweet scented herb with crimson flower heads. Excellent for border and cutting. In flower three months.   Price 10 cts.

Myosotis Palustris, Forget-me-not. Light blue flowers, and in bloom the whole summer; very hardy, especially valuable for its remarkable blooming qualities.   Price 15 cts.

Pansy, Bugnot's Superb. This is the most famous of all the fancy strains, being blotched, stained and veined in unsurpassable beauty and variety of coloring. Price, three for 20 cts.; twelve for 60 cts.; seed, 25 cts. per pkt. of about fifty seeds.

Pansy, German Strain. In many of the finest colors; no other strain equals it for great variety of shades and blending of colors. This Pansy blooms very freely. For variety, see collection Pansy seed. Sold only in mixed lots. Price, large plants, twelve for 50 cts.; smaller for mailing, twelve for 25 cts.; twenty-five for 50 cts.

Pansy, Trimardeau. A distinct and beautiful new race, the flowers of which are of very large size, and the plants may be expected to produce many of the beautiful shades of color that are found in other classes of this popular plant. The true "Trimardeau" Pansies of French origin, seed much less freely than any other class. The plants were grown from seed saved exclusively from finely formed flowers of the richest and most varied shades of color. Price, in fine mixed colors, large transplanted plants, 50 cts. per doz., express only; smaller for mailing, 50 cts. per doz.; postpaid; seed, 10 cts.

NOTE.—We include Pansies in hardy plant list, but unless pains are taken they will live but two years.

Pink, Her Majesty. Large white. The best hardy white variety.   Price 10 cts.
Pink, Earle of Carlisle. Variegated in fancy colors. Maroon, rose and white.   Price 15 cts.
Pink, Souv. de Sale. Large pink, one of the best. Very double, fine for cutting.   Price 15 cts.
Pink, Juliette. Best dark sort, quite double, cherry red, variegated pink and white. Price 15 cts.
Pink, May. Flesh pink, beautiful and very fragrant. Splendid for cutting.   Price 15 cts.

NOTE.—The above five pinks are hardy, but it will favor them, in northern states, if slightly protected in winter with evergreen boughs.

Helianthus Rigidus. One of the most desirable of our native varieties, beginning to bloom early in July and continuing until late in fall; flowers single, golden yellow, with dark center. Price 10 cts.

Helianthus Multiflorus, Double Dwarf Sunflower. A double Perennial "Sunflower." It is a great addition to our hardy herbaceous plants. Its dark, golden yellow color is both fashionable and beautiful for cut flowers; grows from three to five feet in height; never fails to flower the first year of planting. Do not be prejudiced against this plant because it is only a "Sunflower." Give it a trial, and you will be surprised and delighted to find how beautiful it is. It is a hardy Perennial, coming up every year, but in extreme North it will need protection or can be taken up like Canna and Dahlias, and kept in cellar. Easily taken for a fine yellow Dahlia, only with ten times the number of blooms.   Price 15 cts.

Heuchera Sanguinea. A floral gem and one of the most valuable additions to the list of hardy flowering plants that has appeared in a long time. The foliage, which is evergreen, is beautifully cut and marbled and is very abundant. The flowers are borne in large, open, clean-stemmed panicles and are of the most clear, bright, cheerful and charming scarlet imaginable. Although the flowers are small, yet they are so numerous in each panicle, and the panicles are thrown up in such profusion, as to produce a most brilliant effect.   Price 25 cts.

Spirea, *Anthony Waterer.* One of the most beautiful hardy plants in cultivation and one which has aroused the greatest enthusiasm among horticulturists everywhere. This new Spirea grows in the form of a low compact shrub, with numerous branchlets, each bearing clusters of charmingly pretty, rose-colored flowers. The plants sent out are strong, well-rooted, and with ordinary care, will flower freely during summer. It will do well in any good garden soil. (See cut.)

Price 25 cts.

Doronicum Excelsum. One of the most effective of the early spring flowering perennials, beginning to bloom in April and continuing at intervals throughout the season, or by planting in pots in the fall it can be readily forced into bloom during the winter months in the conservatory or window garden. Its golden-yellow daisy-like flowers are of large size, being about four inches in diameter.   Price 20 cts.

Helenium Autumnale. Loose rounded heads, nearly two feet through, bright yellow flowers. Strong stems four to five feet high; blooms in August and September. Price 20 cts.; extra strong, 40 cts.

Spirea, *Nana Compacta.* A new large flowering Spirea from Japan. This is a first-class novelty, and we thoroughly recommend it. In garden culture, it flowers freely during the summer, and is perfectly hardy. Its merit consists in compact growth, ample foliage of brilliant green and its wonderfully free production of feathery white flowers borne in plume-like panicles of magnificent porportions; will entirely supersede the old variety, as the individual flowers and spikes are twice the size, and much freer blooming. Fine for winter window garden, if potted in fall and treated same as Hyacinths. (See cut.) Price, clumps, 40 cts.; small roots, prepaid, by mail, 20 cts.

Aquilegia, *Nana Alba* (or Decoration Day flower). This dwarf white Aquilegia or Columbine, is one of the most valuable of recent introduction. Our national holiday in remembrance of deceased soldiers has become a day of floral memory, and the great demand for flowers on May 30 is best known to the florist, and has caused much study to find the best for this season. Price 20 cts.; seed, 10 cts,

Lychnis, *Flos Cuculi Plenissima Semperflorens.* Recently imported from Germany. Of easiest culture, requiring no special care, strong and quick in growth, and almost ever-blooming. Flowers rose color, and in large clusters, having a very light or feathery appearance. It can be planted out in summer, where it continues to bloom till wanted for winter blooms, or it can remain there over winter, it being a hardy perennial. Price 15 cts.

Phlox, *White Lady.* The improvement that has been made in Hardy Garden or Perennial Phlox of recent years is truly wonderful. This variety instead of being a lot of tall, naked stems with small tufts of bloom at the top for a little while in mid-summer as of old, are dwarf and stocky, with flower heads eight to ten inches long and six to eight inches in diameter, and are produced from June until frost. One of the finest of its class. Pure snow white, a most profuse bloomer, and of dwarf, compact growth. Price 20 cts.

Phlox, *Wm. Robinson* (new). Fine truss of immense florets, often larger than a half dollar; color, new, of clear salmon, with a rosy crimson eye. A showy variety and very hardy. Price 20 cts.

Dialetra, *Spectabilis* (Bleeding Heart). One of the best known perennials, with graceful, drooping racemes of heart-shaped flowers of rosy crimson and silvery white. Blooming in spring and early summer. (See cut.) Price, extra strong roots, 30 cts.; small, 10 cts.

Heliopsis Pitcherianus. A desirable hardy herbaceous plant, growing from two to three feet high, and a perpetual bloomer, beginning to flower early in the season and continuing in bloom the entire summer. The flowers are of a beautiful deep golden yellow color, about two inches in diameter, of very thick texture, and are very graceful for cutting, and lasting long when cut; and if left on plant the same blooms will remain four weeks in perfect condition. Price 15 cts.

Yucca Filamentosa, *Adam's Needle.* A tropical-looking plant, with long narrow leaves that remain green the entire year. It throws up a strong flower stem in summer, three or four feet high, bearing a large spike of creamy white, bell-shaped flowers, that retain their beauty a long time; hardy. Price 20 cts,; large clumps, 50 cts.

Gypsophila, *Baby Breath.* A beautiful old-fashioned perennial, possessing a grace not found in any other perennial, and attracting the eye of everyone. When in bloom it forms a symmetrical mass in height, and as much through, of minute pure white flowers, forming a beautiful gauze-like appearance. For cutting purposes it is exquisite, especially in combination with high-colored flowers, and some most lovely effects can be produced with it. Price 15 cts.

Rocket, *Hesperis.* Very pleasing early spring flowering, profuse blooming plants, with fragrant flowers, growing freely in any light rich soil. Flowers pure white and in full bloom on Decoration Day; fine for cutting, a very desirable plant. Price 15 cts.; seed, 5 cts.

Platycodon Grandiflorum. One of the very best perennial plants; in constant flower from early spring to late fall. It forms dense clumps, which are covered with a mass of bell-shaped flowers. Fine for cemetery, for garden or for cutting. Roots are hardy as a Pæony. Price 15 cts.

Hemerocallis Kwanso flore pleno, *Double Crown Day Lily.* A perfectly double flowering sort; a profuse bloomer. The flowers are produced in clusters; they are of a bright orange-yellow, and each petal is marked with a large crimson blotch, forming a circle around the flower, having elegant grassy foliage and handsome, deliciously fragrant flowers. They are perfectly hardy, and thrive in almost any kind of soil, preferring one that is rich, moist and in an open situation. Price 20 cts.

Lily of the Valley. The beautiful delicate white flowers and the peculiar shade of green foliage render this class of plants extremely attractive and desirable, especially so grown out of doors. For house culture, place rather thickly in pots or boxes. Keep in cool place until mid-winter, then bring to window and give plenty of heat and some moisture, and the flowers will soon appear. Orders should not be sent later than May. Our pips (or roots) are of the new improved large flowering sort called the "New Russian." The best of all. (See cut.) Price 5 cts.; 40 cts. per doz.

Pyrethrum Uliginosum, *Giant Daisy.* A grand fall blooming plant, growing five feet high, and covered with large white flowers with yellow centers. Needs a moist, rich soil, and should be in every garden. No hardy plants in our grounds received so much praise as this fine Daisy, growing as it does, from five to seven feet high, and forming in short time a clump from two to five feet across; beautiful as an ornamental plant; also fine for cutting. Price 20 cts.; 3 for 50 cts.

Rudbeckia, *Golden Glow.* See Page 28.

Pæonies. See Page 24.

Iris. See Page 23.

HARDY VINES.

The following are all hardy vines, for permanent planting out doors.

Ampelopsis Veitchi, *Boston Ivy.* An "Ampelopsis" of Japanese origin. It grows rapidly as the old "Virginia Creeper," and attains a height of fifty feet. It clings firmly to any wall, tree, etc. The leaves are small on young plants, which at first are of an olive green-brown color, changing to bright scarlet in the autumn. As the plant acquires age the leaves increase in size. This variety becomes more popular every season, and is without question one of the very best climbing plants for covering brick or stone walls. Hardy except in extreme north. Price, small, 15 cts.; strong plants, 3-year, 40 cts.

Ampelopsis Quinquefolia. *Virginia Creeper.* A very rapid grower, with large dark green foliage, which changes in the fall to scarlet. A very desirable vine for covering arbors, verandas, etc. Extremely hardy. Price, small, 10 cts.; large, 25 to 50 cts.

Cinnamon Vine. A beautiful climber which possesses the rare quality of emitting from its flowers the odor of cinnamon, and very appropriately called the "Cinnamon Vine." The stem dies down every autumn, but grows again so rapidly in the spring as to completely cover any trellis or arbor very early in the season. It is as easily cultivated as the "Madeira Vine," has no insect enemies, and is not affected by drouth; has beautiful heart-shaped leaves, bright green peculiar foliage, and clusters of delicate white flowers, sending forth a delicious cinnamon odor, rendering it one of the most desirable climbers in cultivation. Price 10 cts.; 3 for 25 cts.; 7 for 50 cts.

Chinese Matrimony Vine. A vigorous, hardy climbing plant when trained to an arbor, fastened to a fence, attached to a tree, the pillers of a piazza or in any location where a hardy, vigorous climber is desired. It sends out numerous side branches, so that it covers a great amount of space in a short time, and every new growth is at once covered with small purple flowers which are succeeded by brilliant scarlet berries nearly an inch long, every branch being loaded with them. Price 15 cts.

New Halliana Honeysuckle. A new variety introduced from Japan, and considered a great acquisition; it is ever green, and a constant bloomer; flowers white, changing to yellow; very fragrant; good for trellis or pillar; one of the very best varieties for all purposes. Price 15 cts.

Monthly Fragrant Honeysuckle. A fine, rapid growing variety; flowers large and very fragrant; color red and yellow; a constant bloomer. Price 15 cts.

NOTE. We can furnish, by express, 3-year pot grown Honeysuckles, 50 cts. each.

Dutchman's Pipe, *Aristolochia Sipho.* A vigorous and rapid growing climber, bearing singular brownish colored flowers, resembling in shape a Dutchman's pipe. Its flowers, however, are of little value, beside its fine light green leaves, which are of a very large size, and retain their color from early spring to late fall; perfectly hardy; extra strong plants. Price 50 cts.

Clematis, Paniculata. One of the most beautiful of our hardy flowering vines. The flowers are pure white, and are borne in great panicles or clusters of bloom, fairly covering the plant, so that it is a mass or sheet of fleecy white. These clusters are borne on long slender stems, which stand out boldly from between the glossy green leaves, and the gentlest breeze causes wavy crests of white to flow gently over their surface. The fragrance is delicious, resembling the English Hawthorn blossoms, and is so subtle and penetrating that a large plant in bloom will fill the air with fragrance. It begins to bloom late in August or early in September, at a time when most other hardy climbers are out of bloom, an inestimable advantage. It is a strong, rapid grower, spreading out when trained to wires or strings. It is perfectly hardy in all sections of the country, and we can unhesitatingly say that it is, in our opinion, the most valuable among hardy climbing vines. Price, 2-year, 25 cts.; extra, 3-year, 40 cts.

Clematis, Henryi. Creamy white flowers. A strong grower and very hardy; one of the best of the white varieties; a perpetual bloomer. Price, 2 year, 75 cts.

Clematis, Jackmanii. Flowers, when expanded, are from four to six inches in diameter: intense violet-purple, with a rich velvety appearance, distinctly veined. It flowers continually from July until cut off by frosts. One of the best. Price, 2-year, 75 cts.

Wisteria Sinensis. A magnificent climber, with a rich foliage, and long racemes of very fragrant lilac flowers, which cover the whole plant in May and June; grows rapidly when well established. Price, extra large plants, 50 cts.; small by mail, 20 cts.

HIRAM, MAINE, April 30, 1898.—*Ellis Bros.* Am very much pleased with the plants, which arrived in fine condition. Thank you for the extra seed. Yours truly, MRS. E. W. BOSWORTH.

ORANGE, MASS., Nov. 1, 1898.—*Ellis Bros.* Dear Sirs: Plants are doing finely. I get better satisfaction from your plants than any of any other house.
Yours respectfully, MRS. E. C. BARTLETT.

So. BERWICK, ME., Apr. 27, 1898. *Messrs. Ellis Bros.* Gentlemen: The plants came to hand duly, with the seeds, and were very fine ones. Thank you very much for the good trade you have given me on them. Yours truly, CAROLINE F. TRACEY.

LAWRENCE, MASS., March 6, 1898. *Ellis Bros.:* I am pleased to state that stock came to hand all in good season and in first class condition, in fact, better than any that I ever had come from any other firm. I am at times obliged to buy considerable. I am, yours truly, L. DAVENPORT.

ENFIELD, N. H., Oct. 1, 1898.—*Ellis Bros.* Dear Sirs: Many thanks for the prompt and very satisfactory manner in which you filled my order for plants, which arrived in fine condition.
Very truly yours, MRS. A. C. TRAIN.

Flower Seed.

Our address books show customers that have steadily purchased their seed of us for ten, fifteen and twenty years. Surely this is a better recommendation than any puffy words of ours for the quality of the stock we sell. We are situated in northern New England. Seeds (and plants) which we can call hardy will prove so in nearly the whole country. It has also been proved that as a rule, northern stock can be carried south with improved results, while the reverse is the case when southern stock is brought north. We would also call attention to the fact that with our greenhouse facilities, we are enabled to test the germinating qualities of each variety, so that we know before placing any in packets that it will under fair conditions give to the buyer perfect satisfaction. To the many who will see our seed list for the first time, we would mention that the seed business is no new experiment, that we have dealt in seeds for twenty-seven years, and have a very large list of patrons over the entire United States and Canada. We would respectfully request a trial order.

HOW TO SOW FLOWER SEED

A shallow box is best to sow flower seed in. Make two or three holes in the bottom for drainage and fill nearly to the top with rich loam, *without manure.* Sow the seed evenly. Take two parts of pure sand to one of loam, and mix well. Place a small quantity of this in a fine sieve (it will not hurt your flour sieve if you have no other); do not shake it, but take one hand and rub the dirt through, watching the box, so as to cover it evenly. Small seed, like Petunia, Cineraria, etc., should be but slightly covered, while the larger ones, like Asters, Verbenas, etc., will do better if one-eighth of an inch of this fine covering material is used. Sprinkle after covering. Cut or fold a piece of cloth or paper the size of inside of box, wet, and lay directly on the earth; this should remain till the seedlings are seen breaking through the ground, when it should be taken off, the box placed in a sunny window; not kept too wet, or they will damp off (rot at top of ground). As soon as the second leaves are well out, they should be transplanted into pots or boxes. Small quantities of well decayed manure should then be used. Small seed sown in open ground should be covered but one-fourth to one-half inch deep; sprinkle surface well after covering seed. Then cover entire surface of bed with old hay, grass, newspapers, or any covering material at hand; leave it on bed till young plants begin to break ground, to come up.

COLLECTION PACKETS.

Our collections, if bought in separate packages, each kind or color by itself, would cost customers, at 10 cts. per paper, from $1.00 to $2.50. We know of no other method by which so large a quantity of first-class seed can be sold for so small an amount of money. Each year we have sold more of these collections than on the preceding season, and at this time our sale warrants us in the assertion that there is not a more popular form of selling choice seed now in use in this country. Large quantities are sold to amateurs in every state, also in Canada, and besides this their reputation has brought us many florists as customers. Immense quantities of poorer seed are sold, *but none better.* New customers are requested to give this class of seed a trial.

COLLECTION PANSIES.

In variety, quality and price, we in confidence place this collection in test with any stock sold in this country, We have sold this grade for over twenty years with steadily increasing sales each season. Customers will find it better than ever this year.

Pres. Carnot, white, each petal blotched violet; *Purplish Violet*, large flowers, fine; *Rosy Lilac*, new color, distinct; *Silver Edge*, purple,with white edge; *Victoria*, new red, fine; *King of Blacks*, darkest variety; *Brown-Red*, various shades; *Candidissima*, satiny white; *Coquette de Poissy*, new French sort, distinct mauve color; *Emperor Frederick*, dark red edge, shaded lighter red; *Emperor William*, ultramarine blue; *Fawn Color*, beautiful shade; *Fire King*, yellow, upper petals purple; *Gold Margined*, a splendid sort; *Havana Brown*, new shades; *Lord Beaconsfield*, deep violet purple, shading at top nearly white; *Mahogany Color*, an odd color; *Meteor*, bright brown; *Peacock*, blue, with white edge; *Wallflower Brown*, various shades; *Striped and Mottled*, very odd; *Striped Parisian*, light mottled, new; *Trimardeux*, giant varieties, mixed; *Quadricolor*, a fine combination of colors; *Yellow Giant*, large yellow, black eye; *Yellow Gem*, only medium size, but pure yellow, without eye; *Atropurpurea*, dark purple; *Odier*, the five blotched or stained Pansies; *White with Black Eye*, a good bedder; *Cassier's*, a fine strain in beautiful tints.

Fifteen seeds of each of above thirty sorts, 450 in all.   Price 25 cts.

These collections are made up by taking fifteen seeds of each of above thirty sorts ; *all are then put into one packet and sealed.* You will make no mistake if you plant this brand.

NOTE.—We can furnish any of above thirty sorts separate, in packets of from fifty to one hundred seeds, according to kinds wanted, at 10 cts. per packet.

PEACHAM, VT.—*Ellis Bros.* I find your seeds very reliable. Resp'y, MRS. R. B. KINERSON.

DEXTER, MAINE.—*Ellis Bros.* Though I have had a dozen or more catalogues sent me this season, I have waited for yours which I received today, and have made out my order. The seeds sent last spring were fine, as they always have been, and perfectly satisfactory.   MRS. C. H. WYMAN.

CARIBOU, MAINE.—*Ellis Bros.* I will get my order in early. I have tried a number of seed houses, and like yours best for this reason: the seed and plants prove true to name.   MRS. INEZ GARY.

SPRINGFIELD, MASS.—*Ellis Bros.* Dear Sirs: Please send collection package of your Pansy seeds. I have had them for several years past and find none more beautiful.   GRETA LORD.

SWAMPSCOTT, MASS.—*Ellis Bros.* I have had Pansy seed for cold frames and early blooming from many different places for years, and on the whole, I get the best satisfaction from your seeds, and my Pansy bed is the admiration of my friends,   Respectfully,   MRS. C. P. JEFFERS.

COLLECTION TOM THUMB NASTURTIUM.

Crystal Palace Gem, sulphur colored flowers with dark red spots near base of petals; *King of Tom Thumbs*, bluish green foliage, intense scarlet flower, very showy; *Pearl*, creamy white; *Golden King*, deep golden yellow flowers, dark foliage ; *Spotted*, flowers beautifully spotted; *Ruby King*, pink, shaded carmine, contrasting beautifully with its dark foliage; *King Theodore*, bluish green foliage, flowers almost black; *Atropurpureum*, dark crimson; *Empress of India*, splendid new dark-leaved variety, flowers crimson; *Beauty*, fine yellow flushed and shaded vermilion ; *Ladybird*, (new), the ground color of the flower is rich golden yellow, each petal barred with a broad vein of bright ruby crimson; *Aurora* (new), a shade of pink, beautiful. Ten seeds each of above twelve varieties, 120 seeds in all.   Price 25 cts.

NOTE.—We can furnish, when desired, any of above varieties in separate packets (of fifteen to twenty seed). Price 5 cts.

NEW BEDFORD, MASS., August 28, 1898.—*Ellis Bros.* I have been much pleased with your collection of Tom Thumb Nasturtiums. The colors are beautiful and as varied as one could desire.   MRS. N. A. STANLEY.

EAST RIVER, CONN., March, 1897.—*Ellis Bros.* I have a large flower garden and it is simply admired by all who see it. I believe every seed that I have of you grows.   NELLIE M. PARDEE.

LARGE FLOWERING FRINGED PRIMULAS, (Single.)

Golden Feather, beautiful golden yellow foliage, flowers pure white; a charming variety; *Christata*, quite dwarf, with odd crisp mallow-like foliage, flowers semi-double, in different colors; *Rubra Violacea*, very showy, a grand and beautiful sort; *Punctata Elegantissima*, new variety, flowers velvety crimson, edges spotted with white, very distinct; *Striata*, a beautiful striped sort; *Fern Leaf*, in variety, fine; *Coccinea*, flowers of the largest size, of a beautiful brilliant scarlet, with a clear sulphur eye, exquisitely fringed; *Globosa Alba*, a splendid new and improved white variety, heavily fringed, extra fine; *Globosa Kermesina Splendens*, an improved deep crimson sort, deeply fringed, very bright and distinct; *Globosa Carnea*, new variety, very fine; *Punctata Atropurpurea*, a beautiful and showy variety; *Pulcherrima*, white, with rose or lilac center, very fine.

Five seeds each of above—twelve fringed sorts—making the most desirable packet of Primrose offered in the world, and at one-half the price asked by most seedsmen for much inferior stock and a less quantity. Its germinating quality has been fully tested by us. **Price 25 cts.**

NOTE.—We can supply (to those who want certain varieties of colors) any of the above Primulas in separate packets. Each packet contains about twenty-five seeds. **Price 25 cts. each.**

KINGSVILLE, O., Oct. 12, 1898.—*Messrs. Ellis Bros.* Sirs: I want to tell the success we had with your seeds. Last spring we bought a mixed package of Primula seed, from which we now have fifty-one very nice plants and threw away at least four very small ones. Respectfully, MARY A. KINNEAR.

BUFFALO, N. Y.—*Ellis Bros.* Kindly mail me at once collection package of Verbena and single Primrose seeds. Have not had time as yet to carefully examine your new catalogue. There is one white fern leaf Primrose among those grown from seed obtained from you last year, which is a remarkable specimen, both in symmetry of form, size and number of florets. Many of the single Primroses are as large as a half dollar. The florists who have seen it pronounce it a wonder.
With best wishes, I remain very truly yours, W. F. LAKE.

NEW LONDON, N. H.—*Ellis Bros.* Dear Sirs: The collection packet of Primula seed I had of you last year, was just splendid. I believe every seed germinated. I had over thirty fine plants, and they were admired by every one who saw them. Respectfully, MRS. J. S. BOHANAN.

COATICOOK, P. Q.—*Ellis Bros.* Dear Sirs: Primula, Stocks and Asters, I found them most excellent, germinating freely, and giving me very fine plants; the Primulas handsomer than any I ever saw before. MRS. C. D. DYKE.

COLLECTION LARGE FLOWERED AND FANCY PHLOX DRUMMONDII.

Alba, magnificent large white flower; *Alba Oculata*, white with colored eye; *Atropurpurea*, deep blood purple; *Atropurpurea Alba Oculata*, blood purple with large eye; *Atropurpurea Striata*, blood purple, striped; *Chamois Rose*, shades of light pink; *Coccinea Striata*, scarlet striped; *Isabellina*, straw color, flowers of extra size and fine form; *Kermesina Splendens*, vivid red, very showy; *Kermesina Striata*, vivid red, striped; *Lepoldi*, deep pink, white eye; *Quadricolor Rosea*, rose color, shaded; *Quadricolor Violeacea*, violet color, shaded; *Rosea Striata*, rose color, striped, very fine; *Rosea Alba Oculata*, rose color, white eye; *Stellata*, beautiful sorts with large star in center; *Violacea Alba Oculata*, violet, with large white eye, very fine; *Brilliant*, named for its bright color; *Cuspidata, New Star Phlox*, long points to petals, very odd; *Fimbriata*, new, extra fine, with toothed or fringed petals.

Twenty seeds each of above twenty varieties of large flowering and fancy Phloxes, 400 seeds in all. **Price 25 cts.**

Separate colors of above in packets of from fifty to 100 seeds. **Price 10 cts.**

NOTE.—With the exception of two last fancy sorts, the above named Phloxes are all of the "Grandiflora" or large flowered sort, and it is with pride that we can send this fine collection to our customers of this season; its equal will not be sent out by any other firm, as we are bound to place our collection Phlox the equal in popular favor with our Pansies, Asters and Primulas, as we have studied to get into one packet the finest set of colors the world affords.

7

COLLECTION BALSAM.

White, Light Lemon, Flesh Color, Rosy Buff, Pomegranate, Red, Purple, Lilac, Violet, Tricolor, Rose Spotted, Pure Scarlet, Striped, Solferino, Atrosanguinea, Blood Red, White Tinged Rose. Five seeds each of above sixteen sorts, eighty in all. A fine collection.   Price 25 cts.

CONCORD, N. H.—*Ellis Bros.* Gentlemen: I wish to tell you how much I like your collection of Balsams. I have had seeds from Farquhars, Henderson, John Lewis Childs and others, but I never had such beautiful Balsams as I raised last year from one of your collection packets. They are simply beautiful, and were admired by all who saw them.   Respectfully,   Mrs. W. J. G.

COLLECTION ASTER.

Truffaut's Perfection. Large Flowered, Snow White, Rose, Dark Blood Red, Light Blue, Black-Blue, Glowing Dark Crimson, White and Black-Blue, Rose and White, Carmine, Carmine and White, Crimson Ball, Dark Crimson and White, Dark Red and White, Victoria Red, Lilac Red and White, Violet, Dark Scarlet and White, Sky Blue and White, Black-Blue and White, Light Blue and White. The finest collection of Asters ever offered. Ten seeds each of above twenty varieties, two hundred in all. We shall this season include in our collection, a few seeds each, White Comet, Pink Comet, Semple's White and Mary Semple Asters.   Price 25 cts.

PENACOOK, N. H., March 1, 1897.—*Ellis Bros.* Dear Sirs: I must not fail to speak of the Asters which I raised from seeds ordered from you last year. They were the most beautiful Asters I ever saw and were greatly admired by all. I consider your Asters the finest strain I ever saw without exception.
                                    Respectfully,   MRS. JOHN S. BOUTELLE.

SOMERSWORTH, N. H., May 13, 1897.—*Ellis Bros* Sirs: I have sent you one or more orders every year for twelve years. My garden is just beautiful from seed ordered from you. Asters are my special hobby, and such a display as I have from your seed.   MRS. E. C. HOLMES.

LACONIA, N. H., Feb. 14, 1897.—*Ellis Bros.* The Asters I raised from seeds received from you last year, were the finest I have ever received.   MRS. MARY I. HALL.

AYER, MASS., Feb. 22, 1897.—*Messrs. Ellis Bros.* Please find enclosed sixty (60) cents in stamps. My Asters last year were the most beautiful ones I ever raised. You will know by my order my friends all admired them too.   MRS. M. J. BROWN.

ADAMS, MASS.—*Ellis Bros.* Gentlemen: The seeds I had of you last year were the best and cheapest collections of seeds I ever bought of any firm at any price. I have bought a great many collections of Asters, but I have never had such a large number of plants (over 200), or such a large variety as I had from the packet I bought of you. When I say this, it means a good deal. Asters are my special hobby.
                                    Very truly yours,   F. D. BROWN.

ANDOVER, N. H., Feb. 26, 1896.—*Ellis Bros.* The Asters bought of you were the finest I ever raised; took first premium at New Hampshire State Grange Fair last year and the year before.
                                    MRS. WILLIAM MORRILL.

COLLECTION DIANTHUS.

Diadem, mottled, extra; *Hedewigi Hybrids,* great variety; *Atropurpurea,* fine dark shades; *Laciniatus,* fringed; *Striatus,* striped varieties; *Albus,* new, white; *Mourning Pink,* a new sort, very beautiful; *Imperialis,* fine double, mixed; *Laciniatus Purpurea,* a fine variety; *Laciniatus Capreo,* dwarf, double, beautiful.

The above ten varieties are the finest of all Dianthus; forty seeds of each, four hundred in all, price 25 cts., or any one of ten varieties in packets of about one hundred seeds, price 10 cts.

CONCORD, April 3, 1896.— *Ellis Bros.:* The flowers from Collection Dianthus were so lovely that I could not spare a flower. I have bought seeds of you for several years, and I find that in all varieties you have the best. You may use my name if you care to, for "Honor to whom honor is due," and I always have better success with your seeds than any others.   Cordially yours,   MRS. W. J. M. GATES.

COLLECTION VERBENAS.

Mammoth Red, finest large sorts in shades of red; *Defiance,* medium size florets, fiery scarlet; *Sea Foam,* pure white in fine large trusses; *Mammoth Pink,* beautiful shades of pink, extra large trusses; *Coerula,* medium size florets in shades of blue; *Italian Striped and Large Eyed* sorts, mixed, a grand mixture. Twenty-five seeds of each of above six varieties, one hundred and fifty seeds in all.   Price 25 cts.

COLLECTION TEN WEEKS' STOCK.

New. Large Flowering Giant Stocks, in following twelve beautiful colors: Sulphur Yellow, Fiery Crimson, Brick Red, Light Blue, Brown-Violet, Ash Gray, Dark Brown, Old Rose, Mauve, Purple, White, Blush.

A very large per cent of the above will come in extra large double flowers. Twenty-five seeds each of above twelve varieties. Price 25 cts.

COLLECTION PRIMULA (Double).

Finest double fringed varieties. *Fimbriata Alba.* fl. pl., fine double white, fringed; *Fimbriata Ker.* fl. pl., fine double crimson; *Fimbriata Striata,* fl. pl., double, striped, various shades; *Fimbriata Atropurpurea,* fl. pl., fine double dark shades. *Crimson King,* new: *Fairy Queen,* a charming blush.

Price, five seeds each, thirty in all, 50 cts.

PETUNIA GRANDIFLORA HYBRIDS.

Large Flowering Section in ten finest named varieties; also, *The Fringed Section* in five varieties. The flowers of both sections are of immense size and of fine color, far superior to the ordinary Petunia. Twenty seeds each of fifteen kinds, three hundred in all. Price 25 cts.

DETROIT, MICH.—*Ellis Bros.:* Your Petunias were the finest we had, large, healthy plants, and large and finely fringed with velvety lustrous surface and exquisite colorings. Yours, HORACE GREG.

SWEET PEAS.

Our list contains the best varieties in the different shades and color. All seed sold by us has been grown by one of the most expert Sweet Pea growers in the world, and in a location specially adapted to growing the best stock.

Stella Morse. The blossom opens with quite a suggestion of yellow—a most fascinating shade of deep cream—which becomes a little lighter as the flower grows older. About the third day it is a delicate primrose with a faint rose-pink on the edge, which casts a pleasing blush tint to a mass of the blooms, without destroying the creamy-yellow effect. It has the most perfect form, being of the largest grandiflora hooded type and grows on long stems, with three and four flowers to the stem.

Price, 1-8 oz. packets, 5 cts.

Earliest of All. Everyone will be eager to have this distinct new strain, when we state that it is identical with the Extra Early Blanche Ferry, except that it is more dwarf in growth and comes into full bloom at least ten days earlier. Price, packet, 10 cts.

New Countess. Flowers are of the largest size and are a pure light lavender throughout—both on standard and wings. Nothing more beautiful than a bouquet of these dainty light flowers. Price 5 cts.

Aurora. The flowers are of fine substance, full expanded form, and are truly gigantic in size. The immense flowers are borne three and four on a stem; the stems are extra long and strong. The color effect is gorgeous; both standard and wings are flaked and striped on a white ground, with bright orange-salmon. Price 5 cts.

Ramona. Grand flowers of improved hooded form and large size, rivaling the very largest, both in size and form. Its coloring is delicately beautiful; a creamy-white, daintily splashed and striped with pale pink on both wings and standard. A vigorous grower, usually bearing four blossoms to stem. Price 5 cts.; oz., 10 cts.

Gray Friar. This is decidedly gray in color, unlike any other Sweet Pea in cultivation. The flowers are very large, the light gray color making a most distinct and pretty effect, both on the vines and as cut flowers.   Price 5 cts.; oz., 10 cts.

Daybreak. In color it has a white ground and on the reverse of the standard is a crimson-scarlet cloud, which shows through in the fine veins and network, the outer margins being white. The wings should be white, but are sometimes slightly flaked with crimson. A thrifty grower and free bloomer.   Price 5 cts.; oz., 10 cts.

Catherine Tracy (new). Its form is perfect. The large standard is perfectly round, very thick, expanded and nearly flat. The color is a soft but brilliant pink of precisely the same shade in wings and standard, retaining its brilliancy to the last, burning and fading less than any other pink variety. Its large open flower, soft but brilliant color, will always make it a favorite.   Price 5 cts.; oz., 10 cts.

Double Sweet Peas. From thirty to fifty per cent. of these will come true, i. e., with extra petals. Not any more beautiful than the singles, but as some of our customers may wish to give them a trial, we offer a mixture equal to that offered by other seedsmen.   Price 5 cts.; oz., 15 cts.

Cupid. The foliage is very dark green. The plant does not grow over five to eight inches high, and never more than twelve to fifteen inches in diameter; color pure white.   Price 5 cts.; oz., 10 cts.

Dorothy Tennant. Flowers warm violet, very large and finely formed. Standard broad, incurved or hooded; wings very large, rounded. Price 5 cts.; oz., 10 cts.

Pink Cupid. Dwarf pink and white, similar to "White Cupid," except the color. The seed germinates much better than the white.   Price 5 cts.

Primrose. Very distinct, and the nearest approach to yellow found in Sweet Peas. The entire flower being of a pale primrose-yellow color, large, of good substance and finely formed.   Price 5 cts.; oz., 10 cts.

Mrs. Gladstone. One of the most delicately shaded varieties in our entire list; when opening, the blooms are buff and light pink, changing to a beautiful soft blush, darker at base of petals.   Price 5 cts.; oz., 10 cts.

Miss Blanche Ferry (or Improved Painted Lady). It bears large pink and white flowers like "Painted Lady," but is much more free-flowering, and ten days earlier in blooming. The flowers have a deeper, richer coloring and greater fragrance. Earlier, more and finer flowers, and longer in bloom.   Price 5 cts.; oz., 10 cts.

Boreatton. Very large flowers, borne in threes upon long stems; the color is a fine deep maroon throughout; nothing like it.   Price 5 cts.; oz., 10 cts.

Emily Henderson. Absolutely pure white. The flowers are extra large, and in form, perfection. The stems are stiff and long, giving an added value for cutting. In earliness and continued bloom, it outrivals all competitors, flowering early and continues a veritable "cut-and-come-again" to the end of Autumn. Fragrance, delicious. The best white.   Price 5 cts.; oz., 10 cts.

The Senator. Standards splendidly expanded, color chocolate, shaded and striped creamy white.   Price 5 cts.; oz., 10 cts.

Lottie Eckford. Lovely, long-stemmed flowers, borne profusely in clusters of three; both standard and wings are clear white, delicately shaded porcelain blue, distinctly and broadly margined lavender. Price 5 cts.; oz., 10 cts.

Captain of the Blues. One of the largest blue flowered sorts. Standard very broad and bright purple blue; wings broad, expanded and a lighter and brighter blue than the standard. Price 5 cts.; oz., 10 cts.

Countess of Radnor. Flowers medium sized; standard broad, waved at edge, pale lilac, shaded mauve; wings pale lilac. A distinct and beautiful variety. One of Eckford's best. Price 5 cts.; oz., 10 cts.

Monarch. Splendid large flowers; standards bronzy crimson, wings, a rich, deep blue; three flowers on a stem. Price 5 cts.; oz., 10 cts.

Adonis. New, carmine-rose; a lovely shade. Price 5 cts.; oz., 10 cts.

Princess of Wales. A lovely variety, shaded and stained with mauve on a white ground in a most pleasing manner. Price 5 cts.; oz., 10 cts.

Butterfly. White ground, laced with lavender blue. Price 5 cts.; oz., 10 cts.

Painted Lady. Well known pink and white variety. Price 5 cts.; oz., 10 cts.

Invincible Carmine. A splendid robust growing variety, producing a great profusion of bright, glowing carmine flowers. Price 5 cts.; oz., 10 cts.

Scarlet Invincible. Remarkably fragrant, with bright scarlet-crimson flowers. Best scarlet. Price 5 cts.; oz., 10 cts.

Ellis Bros.' New Mixed. Our own mixture of choicest of above with many other new varieties introduced by Mr. Eckford, and known as "Eckford's Newest Mixture," all combining to make one of the finest assortments, which will surely please. Price, large packets, 10 cts.; oz., 15 cts.; 3 ozs., 40 cts. Older varieties, but a good mixture. Price 5 cts.; oz., 10 cts.; 3 ozs., 25 cts.

New Introductions of 1899.—We can furnish in original packets of introducer the following new varieties: *Navy Blue*, twelve seeds, 25 cts,: *Sadie Burpee*, twelve seeds, 25 cts.; *Pink Friar*, twenty-four seeds, 15 cts.

Note.—Sweet Peas, great favorites, but never as popular as now. Our sales are very large, which is in part due to our care to sell only *new seed*, grown *for us* in one of the most favorable locations in the world for the growing of this seed. Our list is very select, we offer only the best in the finest and most decided colors, in both new and older sorts. If sown very early, should not be covered more than one-half inch deep.

DESCRIPTIVE LIST OF FLOWER SEEDS.

Aster. *Sinensis.* The single flowering Asters of our grandmothers and great grandmothers, in many fine colors. It is well known that single flowers have more brilliant and purer colors than double ones, and it is not surprising that the taste for them has been steadily increasing. We do not offer these single Asters as anything wonderful, but at this time single flowers are having a great many admirers, in fact the single form of many flowers are much the most graceful. Few except the very oldest people now living have seen the earlier form of this our most popular annual; therefore quite a novelty to the present generation. Price 5 cts.

Fancy Asters. We are enabled, this season, to place within reach of any of our customers, and at moderate expense, the finest assortment of Fancy Asters ever offered in one package. These are selected by us, the fancy shades from the various classes of large flowering Asters, and then combined in one package, over thirty different shades and colors. We name only a few: White, with blood red center; Rose and White, Lilac and White, Light Blue and White, White, with buff center; Crimson and White, White and Dark Lilac, Light Yellow, Light Blue, tipped white;

Cinnabar Crimson and White, White, turning to azure blue; Apple Blossom, Indigo and White, Harlequin striped, assorted colors with white centers; Light Pink and White, and many varieties in fancy shades. We think Aster seed has never been offered by any seedsmen, combining so many shades in one package. It will please customers to be able to procure so large a variety at so small a cost. Price 20 cts.

Aster, *Yellow.* The nearest approach to pure yellow yet found in Asters. Color, pure bright sulphur yellow, the center made up in globe shape, very solid with yellow quills, and bordered with wider petals of a lighter shade.   Price 10 cts.

Aster, *Lady in White.* The leaves of this variety are of a long and very narrow shape, giving to the plants an extremely graceful appearance, and relieves them of that stiffness common to most other classes of Aster. It flowers gradually and so prolongs the blooming period into the late autumn. It is of great value for bedding, for pots, and especially cut flowers and floral work. Pure white.   Price 10 cts.

Aster, *Comet Pink.* A fine early blooming variety, color pink.   Price 10 cts.

Aster, *Comet Blue.* Same as above except in color.   Price 10 cts.

Aster, *Comet White.* A new, mammoth white variety of this popular strain. Flower similar to pink and blue "Comet," except they are larger, more fluffy and graceful, and plants are much larger and stronger in growth. One of the most pleasing of all varieties, either for ornamental beds or cutting.   Price 10 cts.

Aster, *Comet.* Above sorts mixed.   Price 10 cts.

The shape of the flowers in the above new class, differs from all other Asters in cultivation, resembling closely the large flowered Japanese Chrysanthemums. The petals are long, somewhat twisted or wavy-like, are recurved from the center of the flower to outer petals in such a manner as to form a loose, but still dense, half globe.

Aster, *Mignon.* Flowers small to medium in size and pure white; a profuse bloomer. The fine qualities of this Aster have been amply demonstrated by the experience of last season, and it will rank henceforth as one of the most desirable kinds; fine for cutting.   Price, mixed, 10 cts.; pure white, 10 cts.

Aster, *Mary Semple.* New Chrysanthemum-flowered Aster, of a beautiful shade of pale pink, the finest late variety we have ever grown.   Price 10 cts.

Aster, *Semple's White.* Similar to above, only pure white.   Price 10 cts.

Aster, *Semple's Lavender.* A pleasing shade of light lavender.   Price 10 cts.

Aster, *Semple's Crimson.* Fine crimson.   Price 10 cts.

Aster, *Semple's Mixed.* Mixed seed of above four sorts.   Price 10 cts.

The above Asters originated with Mr. James Semple of Pennsylvania, and we pride ourselves in being the first retail seedsmen in the country to offer them to their customers. All are of the branching class, long stems, requiring plenty of room, not nearer than fifteen inches each way, and if ground is good you will have Asters that are the wonder of your friends.

ROUND POND, ME.—*Ellis Bros.* Dear Sirs: I must not fail to speak of the "Semple" Asters which I raised from seed ordered from you last year. They were the most beautiful Asters I ever saw and were greatly admired by all.   Yours truly,   MRS. W. C. THOMPSON.

BELFAST, ME.—*Ellis Bros.* I have sent one order for seed, for my own planting, but my friends want some Asters like those I grew last year, "Mary Semple" and "Semple's White," and so I send again. Our florist does not grow any like them. They are the finest Asters I have grown; last year all of my Asters were a failure, except these.   Yours respectfully,   MRS. GEO. F. RYAN.

Aster, *Princess* (or Snowball). One of the finest white Asters. Resembling a white Liliput Dahlia. In form they are semi-spherical and composed of quite short and very thickly set imbricated petals. A single plant develops as many as thirty pure white flowers, which, by reason of their refined form, may be utilized with the greatest advantage for all purposes for which white flowers are in request; they remain longer in good condition than any others.   Price 10 cts.

Aster, *Truffaut's Perfection.* This improved Perfection Aster is now acknowledged as among the finest for ornamental beds, for cutting purposes or for exhibition, being of strong growth, large bloom, which is extra double, and of fine form. (See also Collection Packet.) The above in following colors separate, viz: Snow white, blood red, pink, and blue. Price, either color, 10 cts.

Aster, *Truffaut's Perfection.* Many colors, extra fine, mixed. Price 10 cts.

Aster, *Truffaut's Perfection.* Good quality, mixed. Price 5 cts.

Aster, *Rose Flowered.* The flowers are large and double, the outer petals finely recurved and the inner ones incurved like a rose; two feet in height; extra choice mixed. A splendid Aster, fine for cutting. Price 10 cts.

Aster, *Victoria Needle.* One of the finest of the quilled Asters. Price 10 cts.

Aster, *Vick's Branching.* A largely advertised white Aster, quite similar to "Semple's White." We offer genuine stock. Price 10 cts.

Aster, *New Victoria* (Large Flowered). It is impossible to speak too highly of this magnificently imbricated Aster. The blossoms are large and distinguished by an elegant and regular overlapping of the petals, thus giving to the flowers a distinctive character. The growth is an elegant pyramid, and each plant grows from twenty to forty flowers. The colors include many extremely delicate and some gorgeous shades. Also variegated. Price, finest mixed, 15 cts.

Aster, *Dwarf Chrysanthemum-Flowered.* A splendid race of dwarf, compact habit, nine inches in height; flowers large and produced when other varieties are out of bloom; mixed. Price 5 cts.

Aster, *German Quilled.* Double quilled flowers, mixed colors. Price 5 cts.

Aster, *Dwarf Pyramidial Bouquet.* Grows only about one foot high, growing in bouquet shape; mixed colors. Price 10 cts.

Aster. All sorts mixed. This will suit the children. Many kinds, lots of colors, large packages, good and cheap. Price 5 cts.

NOTE.—We would call special attention to our Collection Asters, Fancy Asters, Comet Asters; also the Semple's Asters, which need no recommendation to old customers. They are best of all the late Asters. New customers should give them a trial, as they will give satisfaction where any Aster will; all new seed germinating quickly and giving great pleasure and value. No better stock can be found in this country.

Abutilon. Flowers freely during the spring and winter months in the house, and during summer when bedded out; the flowers are bell-shaped; mixed. Price 10 cts.

Aquelegia, *Chrysantha* (Golden Spurred Columbine). A beautiful variety; flowers bright yellow; produced freely all summer; very hardy. Price 10 cts.

Aquelegia, *Columbine.* Hardy perennials, extra quality, mixed. Price 10 cts.

Aquelegia, *Chrysantha Nana Alba* (new). Early blooming, dwarf white Columbine or "Decoration Day Flower." See description, page 43. Price 10 cts.

Ampelopsis, *Veitchii.* The beautiful Japan Ivy: dark green leaves changing to brilliant scarlet in the autumn; hardy, and clings tenaciously. Price 10 cts.

Alyssum, *New Dwarf Sweet* (Little Gem). Plants are of very dwarf, compact, spreading habit, and only three to five inches in height, each plant covering a circle from fifteen to thirty inches in diameter. It begins to bloom when quite small. The plants are a solid mass of white from spring to late in autumn. Price 5 cts.

Alyssum, *Sweet.* Too well known to need description. Price 5 cts.

Auricular, *or Hardy Primula.* Our seed of this grand perennial plant is from one of the best English collections. Flowers are of various shades, yellow, crimson and maroon. Takes a long time for seed to germinate. Price 25 cts.

Antirrhinum, *Queen of the North* (new). In this new sort we have the finest and most beautiful of all Snapdragons. The plants grow into handsome regular bushes, twelve inches high, covered with large, white flowers of a deliciously sweet perfume. Suitable for bedding and pot culture.   Price 5 cts.

Antirrhinum. Mammoth white and shades of cream, mixed; very fine for cutting.   Price 10 cts.

Begonia, *Tuberous Rooted.* These splendid varieties are covered the whole summer with bright and elegant drooping flowers. Blooming the first season from seed, if sown in February or March. Seed is very fine, and great care is needed in sowing and caring for young plants.   Price 25 cts.

Begonia. Fine mixed, from our own fine collection of fancy leaf and flowering kinds; extra.   Price 15 cts.

Coreopsis Lanceolata, *Perennial Golden Coreopsis.* Flowers are each two to three inches or more in diameter, of an intensely clear, golden yellow. It commences to bloom in June, continuing until hard frosts.   Price 10 cts.

Carnation, *Hardy Garden.* From flowers of excellent quality. The finest mixed, hardy garden Pinks; will live in open ground without protection.   Price 25 cts.

Carnation, *Grenadin.* White. A new hardy, dwarf sort; very fine for garden culture; a profuse bloomer.   Price 25 cts.

Carnation, *Grenadin.* Red, same as above, except color.   Price 15 cts.

Carnation, *Mad. Guilland.* In this, we have the best yellow Carnation that will bloom from seed in from four to six months. The flowers are very large and double, borne on long, stiff stems, and clear golden yellow.   Price 25 cts.

Carnation, *Chabaud's Double Perpetual.* This new Carnation grows from sixteen to twenty inches high, is closely branched, and carries its blooms well. Flowerstems are covered with large, handsome, very double flowers in every variety of color. It blooms in about seven months after being sown, and continues to flower in the greatest profusion indefinitely.   Price 25 cts.

Carnation, *Marguerite* (Improved). A new class of Carnations that are without exception the most abundant bloomers of all the Pinks. The flowers are of brilliant colors, ranging through many beautiful shades of reds, pinks, white, variegations, etc.; they are of perfect form and large size. They bloom in about four months after sowing the seeds; those sown in spring commence flowering in early summer, and continue to bloom in profusion until checked by frost. They can be potted and taken in the house and will flower through the winter. They come from fifty to eighty per cent. double.   Price, finest mixed, 10 cts.; pure white, 10 cts.

Carnation, *Perpetual or Tree.* Seed saved from the collection of the celebrated M. Alegatiere, the most celebrated grower in the world, and from finest named stage flowers only; extra for pot culture.   Price 50 cts.

Calendula, *Oriole.* This is surpassingly grand and brilliant. The extra large double flowers are rich and glowing in tone, bright golden yellow.   Price 5 cts.

Calceolaria, *Hybrida Tigrida.* Spotted. Seed saved from the best collection in Europe; extra choice.   Price 25 cts.

Cineraria, *Hybrida Grandiflora.* Saved only from extra fine, large flowering, prize varieties. This strain is unsurpassed.   Price 25 cts.

Cyclamen, *Persicum.* Charming, bulbous rooted plants, with beautiful foliage; winter and spring blooming. If the seed are sown early in the spring, they make flowering bulbs in one season.   Price 25 cts.

Cyclamen, *Giganteum.* Without exception, the strain offered is in all respects the finest that can be procured. The flowers are of large size and of the finest shades.

We can safely recommend this to our customers as sure to give satisfactory results. Extra mixed. Price 50 cts.

Campanula. A wonderful fine class of hardy perennials well suited to any kind of soil. Seed from the most select varieties only. Finest mixed. Price 10 cts.

Campanula, *Persicæfolia* (Angel Bells). See description, page 42. Price 10 cts.

Candytuft, *Fragrant White.* White, fragrant; fine for bouquets. Price 5 cts.

Candytuft, *Empress.* Produces large trusses of pure white flowers; assumes, when in full bloom, a beautiful tree form; very fine. Price 5 cts.

Candytuft, *Rocket.* Large umbels of pure white flowers. Price 5 cts.

Candytuft, *Dark Red.* Fine for cutting, a strong grower. Price 5 cts.

Candytuft. New, dwarf hybrids, mixed; new and very effective. Price 5 cts.

Candytuft. All sorts mixed. Price 5 cts.

Cosmos, *California Hybrids.* New hybrids now introduced for first time. Some of the flowers in this new strain measure five inches across; some flowers are as round as a cart-wheel, with broadly overlapping petals. Some have petals smooth, flat and waxy, others are pleated and frilled at the edges, others toothed and fringed. Some have only five petals, forming a perfect star. There are so many shades of color and such a variety of forms that it is impossible to describe. A great improvement on other strains; but not as early as following. Price 10 cts.

Cosmos, *Dawn.* This new variety comes into full flower in July or August, and continues a mass of bloom until cut down by severe frost; very dwarf, compact growth; flowers are large and a beautiful white, relieved by a delicate tint of rose at the base of the petals. Price, small packets, 10 cts.

Cosmos, *Early Flowering.* A few years ago the Cosmos bloomed with the Chrysanthemum, and being rather more tender, often died an untimely death by frost just at its first blossoming, so that it was almost useless to plant it in the Northern sections, but since this early flowering strain has been produced it may now be had in bloom from July to November. It is such a strong, vigorous grower and its fine fringe-like foliage is such a pretty background for its lovely spreading flowers of white, pink, yellow and rosy purple, that it can hardly be spared. from any garden. It grows with the greatest freedom in any soil, but repays right royally the best care you can give it. One of the most satisfactory of annuals; fine mixed. Price 10 cts.

Chrysanthemum, *Frutescens.* The White Marguerite, or Paris Daisy. Can be grown by any one, producing quantities of white flowers. Price 10 cts.

Chrysanthemum, *Anthemis Coronaria, fl., pl.* A most useful plant for bedding or pot culture. It bears double flowers profusely during the season, and can be highly recommended. Better known as Double Marguerites. Comes quite true from seed. Either white, yellow or mixed. Price, each 10 cts.

Chrysanthemum. Showy and effective garden favorites. Also now extensively grown for cut flowers. No place is complete without them. The annual sorts bloom the first season. The perennial varieties are very fine for out-door bloom, or for pot culture, blooming in October and November. The seed will produce single, semi-double and double flowers in great variety of colors. Price, finest mixed annual sorts, 5 cts.; fine perennial sorts, 10 cts.; finest perennial sorts, extra, 20 cts.

Cobea, *Scandens.* A climber of rapid growth, flowering the first season if sown early in the house or hot-bed. In sowing, place seeds edgewise and cover it with light soil. Purple flower. Price 10 cts.

Cobea, *Alba* (new). Similar to above, except of pure white color. Price 20 cts.

Cyperus Alternifolius, *Umbrella Palm.* A fine ornamental plant. Price 10 cts.

Centaurea, *Marguerite.* Named after Queen Marguerite of Italy, who is so well

8

beloved by the people of that sunny land. This novelty is entirely distinct from all other Centaureas and the most beautiful variety known. The large flowers are of purest white, also in tints of lavender, pink, and lemon, deliciously scented, exquisitely laciniated, and freely produced on long stems, which render them valuable for cutting. It is entirely unlike any other flowering plant.   Price 10 cts.

Daisy, *Mixed* (Bellis Perennis). A favorite plant for beds or pot culture, in bloom from April to June; seed from the finest double varieties, mixed colors; half-hardy perennial.   Price 10 cts.

Daisy, *Double White.* Extra double, pure white Daisy.   Price 10 cts.

Dictamus Fraxinella. A showy perennial, forming a bush two feet in height, having fragrant foliage, and spikes of curious flowers giving off during hot weather a fragrant volatile oil, which explodes when a match is applied.   Price 10 cts.

Dahlia, *Single.* Many of the varieties of these Single Dahlias are exceedingly beautiful, and the seed we offer is saved from one of the best collections, and may be expected to produce many distinct, desirable sorts.   Price 10 cts.

Dahlia, *Double.* Seed of this strain was saved from the very finest double flowers only, and will produce flowers equal, if not superior, to any in the market. An extra mixed strain imported from Germany.   Price 15 cts.

Delphinium. Finest hybrids of the double flowering perennial hardy Larkspurs. Seed saved from named varieties only. Extra quality.   Price 15 cts.

Dolichos. *Hyacinth Bean.* A beautiful class of quick-growing, ornamental climbers, producing an abundance of clustered spikes of flowers, followed by exceedingly ornamental seed-pods : white, purple or mixed.   Price 5 cts. each.

Gypsophila. The annual sort of "Baby Breath." A delicate, free-flowering little plant, covered with a profusion of star-shaped blossoms. Well adapted for hanging baskets and edgings; valuable for making bouquets.   Price 5 cts.

Gloxinia, *Emperor Frederick.* The large and erect flowers, averaging three inches across, are of a fiery scarlet color to nearly the base of the throat, while the margin of the petals is occupied by a clearly defined, pure white band. Come true from seed.   Price 25 cts.

Gloxinia, *Hybrida Alba.* New pure white variety, fine flowers with soft yellow throat ; a beautiful and rare sort, coming true from seed. One of the finest house plants offered in this catalogue.   Price 25 cts.

Gloxinia, *Crassifolia.* Varieties mixed. Strong growing kind ; splendid large finely shaped flowers, in beautiful colors.   Price 25 cts.

Gloxinia, *Dark Blue.* True from seed. Very large flowers ; a dark velvety blue ; extra fine.   Price 25 cts.

Gaillardia, *Picta Lorenziana.* A profuse flowering, double variety, fine for massing, and useful as a bouquet flower, blooming until frost ; mixed. Price 10 cts.

Hollyhock, *Mammoth Fringed Allegheny.* An entirely new departure which has much to commend it. The mammoth flowers are wonderfully formed of loosely arranged fringed petals, which look as if made from the finest China silk, and when cut are hardly recognizable as a Hollyhock, having none of the formality of the ordinary type. The color varies from the palest shrimp-pink to deep red. The plants are of majestic growth, sending up spikes six to seven feet high, and are of a strong constitution, not being troubled with the disease so prevalent among Hollyhocks.   Price, small packet, 10 cts.

Hollyhock. The Hollyhock in its present state of perfection, is very unlike its parent of olden time. It now ranks with the Dahlia, Aster, Camelia, etc. The flowers are as double as a rose, of many shades of color from deep yellow, red, purple, to white. Hardy perennial, three to five feet. Our seed is from the celebrated Chaters strain. No better can be found. Finest mixed. Price 10 cts.

Heliotrope. Fine mixture of older sorts. Price 10 cts.

Heliotrope, *Lemoines Giant Hybrids.* New varieties with immense heads of blooms. In open ground these have often measured twelve to fifteen inches across. Color from almost white to deep indigo. Very fragrant. This novelty comes from France. Price 15 cts.

Humulus, *Japonicus.* (Japanese Hop.) Same as the following, except with plain green leaf; a fine summer climber. Price 5 cts.

Humulus Japonicus, *Variegated.* It is an annual, a very rapid grower, useful and ornamental. The leaves,—averaging from six to seven inches across—are beautifully and distinctly marked with silvery white, yellowish green and dark green, partly regularly striped, as well as marbled and blotched. It is not injured by insects, does not suffer from the heat. This plant will rapidly cover porches, fences, summer houses and rustic arches. Three-fourths plants come variegated. Price 10 cts.

Lathyrus (Perennial Peas). The Perennial Peas, while lacking the delicious fragrance of the Sweet Peas, are equally as beautiful, and are very valuable from the fact that they are in full bloom early in the spring, before the others have begun to flower. They live year after year, bearing magnificent clusters of flowers. Perfectly hardy and grow six to eight feet high. Not new, but not as well known as should be.

Lathyrus, *Purple.* Or dark rose. Price 5 cts.

Lathyrus, *Splendens.* Dark red, fine cut foliage. Price 10 cts.

Lathyrus, *Pure White.* Fine for cutting. Extra. Price 10 cts.

Lathyrus, *Mixed.* Of above. Price 10 cts.

Morning Glory, *Imperial Japanese,* (new). These grand new Morning Glories from Japan are remarkable for the large size and exquisite new colors of the flowers, with magnificent foliage, often beautifully blotched. Of strong, robust growth. They are of all shades of red, from the most delicate pink to the most brilliant crimson, maroon, blue, and pale lavender to richest indigo and royal purple; also white, yellow, gray, slate, copper color, brown, bronze, almost black, and many other odd shades. Many have a distinct marginal band of a different color from the rest of the flower, and some are as elegantly spotted; others are striped, blotched, mottled, rayed, and shaded, often having seven or eight colors and tints in one flower. Many are of very odd and singular forms. Price 10 cts.

Morning Glory (Convolulus). Old-fashioned sorts mixed, still one of the best quick growing climbers. Price 5 cts.; oz. 15 cts.

Marigold, *El Dorado* (or Dahlia-flowered). The large, bushy plants, of good habit, are each a ball of brilliant colors, many single plants having seventy-five to one hundred flowers in full bloom at one time. The flowers are globular, as perfectly double as a Dahlia, and measuring three inches across. Price, mixed, 5 cts.

Marigold. Finest dwarf varieties; flowers in profusion, in shades from sulphur color to dark orange; mixed. Price 5 cts.

Marigold, *Prince of Orange.* Beautiful new variety, florets being striped with an intense shade of orange. For effective and persistent blooming in beds and groups it has no superior, its beautiful flowers being produced from early spring until late in the autumn. Price 5 cts.

Mignonette, *Matchet.* Of French origin, and a most desirable variety. The plants are dwarf and vigorous, of pyramidal growth, with very thick, dark green leaves. They throw up stout flower stocks, terminated by long and broad spikes of deliciously scented flowers, of a red tint. Price 5 cts.; oz., 40 cts.

Mignonette, *Defiance* (new). When well grown, the spikes will grow from ten to fifteen inches long. The flowers are of most delicious fragrance. The individual florets are of large size and stand out boldly, forming a graceful as well as compact spike. It also possesses extraordinary keeping qualities, the spikes having kept three weeks after cutting, retaining their grace and fragrance to the last.   Price 10 cts.

Mignonette, *Parsons' White.* Flowers nearly white; a desirable variety; good fragrance.   Price 5 cts.

Mignonette. Large flowering, fragrant varieties.   Price 5 cts.; oz., 15 cts.

Mignonette. Red flowering, or shaded with red tint; fragrant.   Price 5 cts.

Mignonette, *Miles' Spiral.* Entirely distinct from the old varieties. It is a strong grower and abundant bloomer, producing spikes from eight to eleven inches in length; fragrant.   Price 10 cts.

Mimulus, *Tigrinus.* An exceedingly beautiful new blotched and spotted Hybrid. Numerous varieties, with white, sulphur and yellow grounds, spotted crimson, scarlet and pink. They luxuriate in damp, shady situations, blooming freely in early summer months. Extra mixed.   Price 10 cts.

Nicotiana, *Affinis.* An annual, with sweet scented, pure white, star-shaped flowers, three inches across, blooming continually, except in mid-day.   Price 5 cts.

Nasturtium, *Dwarf or Tom Thumb.* In twelve named kinds. See collection packet, page 48.   Price, either sort, 5 cts.

Nasturtium, *Dwarf.* A fine mixture.   Price 5 cts.; oz., 15 cts.

Nasturtium. Tall or running sorts, in following sorts: Dark crimson, pearl, nearly white, scarlet, yellow, orange, straw color spotted, scarlet striped. All fine decided colors.   Price, each, 5 cts.

Nasturtium. Tall, mixed sorts; first quality.   Price 5 cts.; oz., 15 cts.

Nasturtium, *Lobb's.* The leaves and flowers are somewhat smaller than the ordinary tall Nasturtium, but their greater profusion renders them superior for trellises, arbors, for hanging-over vases, rock-work, etc.; the flowers are of unusual brilliancy and richness; splendid for winter decoration, either for hanging baskets or training about windows.   Price, mixed, 5 cts.; oz., 25 cts.

Nasturtium, *Madame Gunter.* These new Hybrids make a strong growth, climbing from five to six feet high, and are covered from spring to fall with large, substantial flowers of the most brilliant shades. They are remarkable for their wide range of colors. Also striped and blotched in the most fanciful manner. This new strain deserves a place in every garden.   Price, packet, 10 cts.; oz., 30 cts.

Nasturtium, *Chameleon* (tall). This highly interesting new climbing variety presents a most unusual feature, which is, that the flowers are of different colors on the same plant, and that they are variously blotched, splashed, striped and and bordered with these different colors. On one and the same plant, self-colored flowers are found, while others are stained and flushed on clear ground, and others have either light or dark margins. The effect of the variously colored flowers on one plant is simply marvelous, and we consider it one of the most distinct and useful novelties of recent years.   Price 10 cts.

Nasturtium, *Vesuvius* (tall). A new and yet rare sort and quite a novelty. Flowers pink-tinted salmon, foliage very dark, splendid.   Price 10 cts.

Nasturtium, *Lilliput.* This new class of Dwarf Nasturtium might be styled a miniature form of the "Lobb's Nasturtium, being evidently a sport from this beautiful climber. The plants and flowers are both smaller than those of the Dwarf Nasturtium and present a most charming appearance. The dainty little flowers are produced in abundant profusion right over the prettily veined leaves. This new mixture contains all the rich velvety and soft color variations of the Nasturtium family, and some new shades. Price 5 cts.

Petunia, *Grandiflora Fimbriata, fl. pl.* Flowers are of immense size and exquisite colors, shade and markings; have beautifully fringed petals, making them very attractive. It is, perhaps, not generally understood that the seeds of double Petunias are only obtained by artificially fecundating single blooms with the double, making the seed necessarily high priced; and the progeny cannot be expected to come all double. Price, per pkt. of about 50 seeds, 50 cts.

Petunia, *California Giant.* A fine strain of the large flowering varieties in both the gorgeous and delicate shades, and many of them giants in size. This strain is very similar to our collection, except not in so large a variety, but some blooms of larger size. Price, about 100 seeds, 25 cts.

Petunia. Collection packet. See page 51. Price 25 cts.

Petunia. Double varieties mixed. Many come striped and blotched; about one-fourth of the plants may be expected to produce double flowers; but the singles from these strains are also fine and desirable. Price 25 cts.

Petunia, *New Dwarf Inimitable* (single). A Petunia which is really dwarf. It forms a compact plant from five to eight inches high, by as much in diameter, and thickly covered with cherry-red flowers, each of which is marked by a white star. Fine bedding sort. Price 5 cts.

Petunia. Single fine mixed, solid color; also, striped and mottled, in a great variety of shades. Price 5 cts.

Platycodon Mareisi, *Hardy Perennial.* A very handsome new variety for garden decoration in the summer and fall. It grows sturdy and compact, its many branches bearing numbers of beautiful, large open, bell-shaped flowers of a rich violet-blue. The appearance of the plant in bloom is exquisite. Price 10 cts.

Platycodon Grandiflorum. A hardy perennial, producing very showy flowers during the whole season. They form large clumps and are fine for planting among shrubbery; mixed colors. Price, mixed, 5 cts.

Pansy, *Buguots.* The most fancy strain in existence, blotched, stained and veined in many fancy combinations; extra. Price 25 cts.

Pansy, *English Faced.* These are the true faced Pansies of years ago, as in the race for novelties these have been partially lost sight of. Not as large as many of the other varieties. Finest mixed. Price 10 cts.

Pansy, *Giant.* New, largest flowering mammoth Pansies in finest assortment. This mixture contains nearly all the largest flowering sorts to date, much superior to Trimardeux in the variety and coloring. Price 25 cts.

Pansy Collection. See page 48. Price 25 cts.

Pansy, *Trimardeaux.* Mammoth varieties, mixed; fine strain. Price 10 cts.

Pansy, *Giant Striped.* The newest, mammoth fancy Pansy. Price 25 cts.

Pansy, *Giant White.* Very large with purple eye, a grand sort. Price 25 cts.

Pansies in thirty named sorts, fifty to one hundred seeds according to variety. See description, (collection Pansies), Page 48. Price, each sort, 10 cts.

Pansy. Extra good, mixed; fine for bedding. Price 10 cts.

Pansy. Good mixed; cheap, but good. Price 5 cts.

Phlox, *Drummondii.* Double yellow; new; of a light yellow or straw color; desirable. Price 10 cts.

Phlox, *Cuspidata and Fimbriata.* The plants are fifteen inches high, of erect habit, and produce flowers with five dentated petals, the center tooth of each being elongated and ending into a point about one-third inch in length, the whole forming a regular star. The beauty of the flowers are still enhanced by the broad white margin appearing along its edges. Finest mixed colors.   Price 10 cts.

Phlox, *Drummondii.* Double white. Not merely a tendency to become double, which is apparent, but on the contrary sixty per cent. at least of the seedings produce a profusion of charming, double and semi-double pure white flowers. Price 10 cts.

Phlox Collection Packet. See page 49.   Price 25 cts.

Phlox Drummondii. In twenty separate sorts, about 100 seeds in a packet. See full description, page 49.   Price, each sort, 10 cts.

Phlox, *Drummondii.* Finest mixed.   Price 5 cts.

Phlox, *Drummondii, Grandiflora,* or large flowering kinds, mixed. Price 10 cts.

Poppy, *Silk or Ghost.* This new strain of single sorts is of superlative excellence, and many beautiful new forms, unknown before, will be found. They are especially strong in shades of terra-cotta and combinations of this with pink and scarlet. Many lovers of flowers have the impression that the single poppies are too fragile to stand picking, but such is not the case. If gathered early in the morning while the dew is still on them and before the rays of the hot sun have affected them, and placed in water, they will last in all their beauty for forty-eight hours.   Price 10 cts.

Poppy, *Ellis Bros'. Fancy Mixed.* This brand contains in large packets all the kinds we advertise, and many other sorts which we have procured especially for this mixture. Our customers will find in it the largest variety of Poppies they have ever seen from one package.   Price 10 cts.

Poppy, *Fayal* (new). The seeds of these Poppies came originally from the Fayal Islands. They are charming dwarf Poppies, double and single; they are like crinkled tissue paper, and are every imaginable Poppy color — deepest red, terra cotta, salmon, soft pink, snowy white, white, pink and gray.   Price 10 cts.

Poppy, *Tulip.* Fourteen inches high and produces large flowers of the most vivid scarlet imaginable; the color being seen even from afar, of such glowing richness as to at once remind one of scarlet Tulips.   Price 5 cts.

Poppy, *Iceland* (Papaver Nudicaule). The fragrant, elegant, crushed-satin-like flowers are produced in never-ceasing succession from the beginning of June to October. Not only are they attractive in the garden, but for elegance in a cut state they are simply unsurpassed, and they last quite a week if cut as soon as open. They flower the first season from seed, though they are hardy herbaceous plants, and will with slight protection live over winter in any of the northern states.   Price 5 cts.

Poppy, *Fairy Blush.* Few Poppies can equal this in beauty, and no others remain nearly so long in flower. The immense globular flowers are perfectly double, and measure from ten to thirteen inches in circumference; the petals are elegantly fringed and pure white, except at the tips, where they are colored rosy cream. Price 5 cts.

Poppy, *The Mikado.* From Japan, and is in form and character essentially a Japanese flower in its quaint, yet artistic beauty. The petals at the base are whole, while the edges are cut and fringed in most complete manner. The color is most attractive, being white at the back, while edges are a crimson-scarlet. Price 5 cts.

Poppy, *Shirley.* The colors are blended in the most beautiful and showy fashion, and include almost every shade from pale rose to the most dazzling crimson-scarlet, blotched and variegated in many styles. Very much pains have been bestowed upon them by the raiser, who has been carefully selecting them for years. In choicest mixture. Single varieties.   Price 5 cts.

Poppy, *Danish Flag.* Flowers brilliant scarlet, with a large silvery white blotch at the base of each petal, forming a white cross on scarlet ground. Plant two feet in height.   Price 10 cts.

Poppy. Carnation flowered; finest mixed; very showy. Price 5 cts.

Poppy. Peony flowered; blooms of immense size; best mixed. Price 5 cts.

Poppy, *Snowdrift.* Peony flowered; pure white. Price 5 cts.

Poppy, *C. King.* Peony flowered; bright scarlet. Price 5 cts.

Poppy, *Oriental.* Single dark scarlet with black spots; a hardy perennial; showy; flowers of immense size. Price 10 cts.

Primula, *Obconica.* See description, first pages of this Catalogue. Price 10 cts.

Primula Collection Packet. See page 49. Price 25 cts.

Primula, *(Primrose).* Not quite as large flowers as in our collection packet, but in fine variety; very free flowering. Price 15 cts.

Ricinus Zanzibarensis. New and distinct. Their gigantic leaves, two to two and one-half feet across, and the great size of the plants surpass any other known Ricinus. We offer it in mixture. One will produce light green leaves; another, coppery brown; another, brownish purple; another, with bronze leaves. Price 10 cts.

Sweet Peas. See page 51.

Salvia Splendens, *Scarlet Sage.* Fine for autumn decorations: growing from two to three feet high, and completely covered with brilliant scarlet flowers. Seeds should be sown early in spring, in house or hot-bed. Price 10 cts.

Salvia, *Compacta.* See description, page 41. Price 10 cts.

Scabiosa, *Snowball.* Beautiful new white variety; extra. Price 5 cts.

Scabiosa, *Aurea.* Light yellow, very nice for cut flowers. Price 5 cts.

Scabiosa, *Black-Purple.* This variety has very fine, rich colored flowers, extra for bouquets, and one of the fashionable flowers in large cities. Price 5 cts.

Scabiosa. Finest sorts in many colors, mixed. Price 5 cts.

Stock, *Giant Perfection or White Giant,* (new). Plants of this variety attain a height of two and one-half feet, and produce long spikes of double flowers, much larger than the ordinary Ten Weeks' Stock. Pure white; very fine. Price 15 cts.

Stock, *Giant Perfection.* Same habits as above; fine mixed colors. Price 15 cts.

Stock, *Princess Alice,* (Cut and Come Again). A fine perpetual blooming Stock, growing about two feet high. If sown early, it commences blooming in June and continues until destroyed by frost. Its most valuable feature is that it produces perfect flowers during September and October, when other varieties sown at the same time have faded. It throws out numerous side branches bearing clusters of very double pure white fragrant blossoms, and is excellent for cutting. Price 10 cts.

Stock, *Large Flowered, Dwarf Pyramidal.* Ten weeks. Pure white; a fine low growing sort; extra large truss and floret; very fragrant. Price 15 cts.

Stock, *Large Flowered or Dwarf Pyramidal.* Ten weeks. Same as above in many colors; mixed. Price 15 cts.

Stock, *Large Flowered Dwarf.* Ten weeks. A fine variety for general purposes, for cutting or for show; free flowering; mixed. Price 10 cts.

All of above Stock seed is German grown, from pot plants only, and will produce a large per cent. of finest double flowers.

Stock. Ten weeks. A fine mixture, in many colors. Price 5 cts.

Torenia. A very fine annual, forming a very splendid plant for vases, hanging baskets, for the house, or for growing out of doors. Covered until late in the season with one mass of bloom. Two varieties, sky blue with spots of dark blue with yellow center, and bright golden yellow with a brownish red throat. Price, mixed, 15 cts.

Verbena, *New Golden Leaved.* An effective Verbena with golden yellow foliage, which strikingly contrasts with the flowers. A valuable acquisition for bedding, as it remains unimpaired until frost. True from seed. Price 10 cts.

Verbena, *Sea Foam* (Candidissima). All that need be said of this fine Verbena is that the flowers are sweet scented; trusses large and beautiful, and borne freely at all times, and snow white, always coming true from seed. Very fine for solid bed on lawn; also, one of the nicest of cut flowers for summer bouquets.   Price 10 ¢ts.

Verbena, *New Mammoth.* The characteristics of this new strain of Verbenas are that when well grown, flower trusses are over nine inches in circumference, while many of the single florets are as large as a twenty-five-cent piece; the plant also has the peculiarity of being more vigorous in growth than the ordinary Verbenas. The colors present the same range as the ordinary type.   Price 10 ¢ts.

Verbena, *Defiance.* Splendid scarlet; quite true from seed.   Price 10 ¢ts.

Verbena, *Italian Striped.* A large per cent. of this variety coming striped and splashed.   Price 10 ¢ts.

Verbena, *Mixed.* Finest mixed, from a large collection. This mixture includes "Sea Foam" and other varieties.   Price 10 ¢ts.

Verbena, *Large Eyed.* In fine mixed colors.   Price 10 ¢ts.

Verbena, *Cærulla.* In shades of blue.   Price 10 ¢ts.

Verbena. A good grade of seed; mixed colors.   Price 5 ¢ts.

Wallflower. Deliciously fragrant garden flowers, blooming early in the spring, with long, conspicuous spikes of beautiful flowers. They should be protected in a cold frame or cellar in the winter, and planted out in May. Half-hardy perennial, finest double mixed, all colors.   Price 10 ¢ts.

Zinnia, *Zebra.* The most beautiful and brilliant selection of double Zinnias we have ever seen. All the flowers produced from the seed we offer, will not be striped, but a large percentage of the plants will be; those that are self-colored will be found brilliant in the extreme.   Price 5 ¢ts.

Zinnia, *Large Flowering Dwarf.* A new dwarf section, quite distinct. The flowers are as large or larger than the old class, and of much better shape, resembling Dahlias in form, and the habit of growth is compact and dwarf, rarely growing over two feet high; mixed.   Price 5 ¢ts.

Zinnia, *Giant Mammoth* (mixed). A new class of Zinnias, differing from the older ones in its robust habit of growth, and the immense size (five or six inches across) of the perfectly formed, very double flowers of various striking colors. The plants rise to a height of three feet, are clothed with luxuriant foliage, and bloom freely during a long period. The luxurious growth and the large bright flowers of this novelty make it valuable for large groups; but it will also be found most effective when planted singly, or as a border plant in small gardens.   Price 5 ¢ts.

Zinnia, *Curled and Crested.* Magnificent variety of colors. From this, by persevering selection and careful culture, we have a new strain of double flowering, curled and crested Zinnias, which eclipse in beauty, beyond all question, any other types of this popular garden annual in existence. The flowers are of perfect form—large, round, full and double, the petals being twisted, curled and crested into the most fantastic contortions and graceful forms, rendering them entirely free from the stiffness which was heretofore characteristic of the family.   Price 5 ¢ts.

Zinnia, *Jacqueminot.* Dark crimson, coming quite true from seed; very showy. Fine for massed beds.   Price 5 ¢ts.

Zinnia, *Large Flowering Dwarf.* White.   Price 5 ¢ts.

Flower Seed, *Annuals.* All kinds, mixed, containing nearly all the varieties we advertise, and many besides. These if sown in one bed, will make what is called the crazy bed. These seeds are put up in large packets, and where a quantity of flowers are wanted, and in the largest variety—for a little money—this packet will surely suit.   Price 10 ¢ts.

Other standard flower seeds so well known that description is unnecessary :

	PER PKT.			PER PKT.
ACACIA, finest mixed,	10	ICE PLANT,		5
ACROLINEUM, everlasting, mixed,	5	LARKSPUR, Dwarf Rocket,		5
AGERATUM, Dwarf White,	5	Tall Stock, flowered,		5
Dwarf Purple,	5	LINUM, (Grandiflorum Rubrum), mixed,		5
AMARANTHUS, Tricolor (Joseph's Coat),	5	LUPINUS, mixed varieties,		5
ANTERRHINUM (Snap Dragon), mixed,	5	LOBELIA, finest mixed,		10
BACHELORS' BUTTON, fine mixed,	5	Speciosa, Dark Blue,		10
BALOON VINE (Cardiospermun),	5	LANTANA, finest mixed,		10
BALSAM, Allen's new double white, extra,	10	MARIGOLD, best French, tall,		5
Finest mixed,	10	MARVEL OF PERU, mixed,		5
Good mixed,	5	MAURANDIA, finest mixed colors,		10
CALLIOPSIS, mixed,	5	MUSK PLANT,		10
CANNA (Indian Shot), mixed, very fine,	5	NASTURTIUM, tall, mixed,		5
New French Varieties, mixed, extra,	10	Dwarf, mixed,		5
CENTRANTHUS, mixed,	5	PANSY, mixed seeds of all sorts,		5
CENTAUREA, white leaved foliage plant,	10	PETUNIA, fine mixed,		5
CLARKIA, finest mixed,	5	PORTULACCA, mixed single,		5
COCKSCOMB, mixed; very fine,	5	Finest double, mixed,		10
COCKSCOMB GLASGOW PRIZE, new; very dwarf,		RODANTHE,		10
with immense crimson combs, extra,	10	RICINUS, Castor Oil Bean, mixed,		5
CONVOLVULUS (Morning Glory), mixed,	5	Bourbenensis, twelve feet, extra,		10
Minor, the dwarf or trailing variety,	5	Gibsonii, Dark Red Foliage,		10
CYPRESS VINE, scarlet,	5	SALIPGLOSSIS, large flowered varieties ;		
White,	5	splendid, all colors mixed,		5
Mixed,	5	SCABIOSA (Morning Bride), mixed,		5
CANTERBURY BELLS, mixed,	5	SENSITIVE PLANT,		5
CANARY BIRD FLOWER, climber,	10	SWEET WILLIAM, fine mixed,		5
CHINESE PINK, finest mixed,	5	Extra strain of double sort, mixed,		10
DATURA WRIGHTII, the best,	5	SMILAX, finest climber,		10
DIANTHUS, Heddewigii, finest mixed,	10	THUNBERGIA, mixed,		10
DIGITALIS (Foxglove), mixed,	5	VERBENA, mixed, fine,		10
ESCHSCHOLTZIA (California Poppy), fine		VINCA, mixed,		5
mixed colors,	5	VIOLET (English), best varieties, mixed ;		
FORGET-ME-NOT, White,	5	seed germinates very slowly,		5
Blue,	5			
Mixed,	5	**ORNAMENTAL GRASSES.**		
FEVERFEW, Double White,	10	AGROSTIS NEBULOSA, hardy annual, one foot,		5
Golden Feather,	5	BRIZA MAXIMA (Large Quaking Grass),		5
GILIA, mixed,	5	BRIZA GRACILIS (Slender Quaking Grass),		5
GLOBE AMARANTH, mixed,	5	BROMUS BRIZÆFORMIS; a fine grass,		5
GERANIUMS, finest mixed,	10	CHLORIS BARBATA; pretty and curious,		5
HELICHRYSUM (Everlasting Flowers), finest,		COIX LACHRYMÆ (Job's Tears),		5
mixed,	5	STIPA PENNATA (Feather Grass); hardy,		5
HELIOTROPE, fine mixed,	10	GYNERUM ARGENTEUM (Pampas Grass),		5

VEGETABLE SEEDS.

Our vegetable seeds comprise the best standard and new sorts. We might offer a much larger list of varieties, but it seems useless, as the kinds offered are of the greatest merit, and, in our estimation, all that can be desired. We have taken the greatest pains to produce seeds of the choicest quality, and can assure our customers that the stock here offered, is, in freshness and purity, second to none in the country. All prepaid at prices given, except as noted.

Asparagus, *Conover's Colossal.* Best of all.   Price 5 ¢ts.; oz., 10 ¢ts.

Beet, *Stinson.* True dark red variety. Excellent for the open garden, this is particularly valuable for forcing, because it produces less foliage than any other variety. Fine table beets of two and a half inches in diameter will have foliage scarcely five inches high, the leaves proper being only three inches long, of a deep, rich red color. The roots are smooth and regular in form; the skin is dark blood-red. The flesh is fine grained and free from woodiness. Of a rich color, the beets cook dark blood-red, remain in fine condition a long time. Price 5 ¢ts.; per oz., 10 ¢ts.

Beet, *Crimson Early.* This Beet is the earliest variety in cultivation, and is ready to market in six weeks from time of sowing. The tops are short and green, and the bulbs of beautiful shape. The skin is deep blood-red in color; the flesh being also very dark, of the finest quality, tender and sweet at all times. The beets are of good size and very uniform, and never become tough and stringy with age. As an early Beet it has no equal, and for late sowing its quick, rapid growth, makes it also valuable.   Price 5 ¢ts.; oz., 10 ¢ts.

Beet, *Edmand's Early Turnip.* This is the most uniformly " thoroughbred " of all Beets; with very small tops, the neat short foliage being of a bronzy red. The beets are of a handsome round shape, very smooth and of good marketable size. It does not grow very large and coarse where room is permitted, as do most varieties of the Turnip Beet.   Price 5 ¢ts.; oz., 10 ¢ts.

Beet, *Egyptian.* Early; fine for summer use; turnip shape.
Price 5 ¢ts.; oz., 10 ¢ts.; 1 lb., 50 ¢ts.

Beet, *Eclipse.* Intense, deep blood-red, small top. Sow late for winter use.
Price 5 ¢ts.; oz., 10 ¢ts.; 1 lb., 60 ¢ts.

Beet, *Dewing's Improved Blood Turnip.* Fine turnip form, very symmetrical in shape, and free from any fibrous roots. Flesh deep blood-red, very tender and sweet; grows uniformly to a good size. Early, and when sown late, desirable for winter.   Price 5 ¢ts.; oz., 10 ¢ts.; ¼ lb., 15 ¢ts.; 1 lb., 50 ¢ts.

Beet, *Swiss Chard* (Silver or Sea-Kale). This variety is grown for its leaves only; the middle of the leaf is cooked and served like Asparagus; the other portions of the leaf are used like Spinach.   Price, pkt., 5 ¢ts.; oz., 10 ¢ts.

Beet, **Mangel Wurzels.** For stock. The three following sorts are among the best: *Norbitan Giant,* improved long red, an immense yielder; *Lane's Imperial Sugar,* not as large as above, but very sweet; *Golden Tankard,* an improved yellow globe, a great yielder. Price, either sort, oz., 5 ¢ts.; 1-4 lb., 15 ¢ts.; lb., 40 ¢ts., prepaid; by express, not prepaid, 30 ¢ts. per lb.

Cabbage, *Succession.* About one week later than the "Early Summer," but it is double the size, and is so finely bred that in a field of ten acres every head was a perfect specimen, fit for exhibition purposes. Its earliness, uniform large size, solidity and sure heading qualities, makes it the most valuable Cabbage either for private use or market purposes.   Price 5 ¢ts.; oz., 15 ¢ts.

Cabbage, *Early Winningstadt.* A well known and very popular second-early variety; heads large, decidedly conical; leaves bright, glossy green; heads very solid and hard, even in summer. This is a very sure heading variety, valuable not only for early use, but also for winter Cabbage; its vigorous growth enables it to withstand unfavorable conditions. Price 5 cts.; oz., 15 cts.

Cabbage, *Burpee's All-Head.* The nearest approach to a "thoroughbred" Cabbage of any variety yet introduced, both as regards size and uniformity of development. Selection has been constantly made to secure the largest early heads, with the fewest loose leaves. The deep, flat heads are solid and uniform in color, as well as in shape and size. By reason of the compact habit of growth, and absence of loose leaves, fully one thousand more heads can be obtained to the acre, than of most other varieties of same size. Price 5 cts.; oz., 15 cts.

Cabbage, *Fottler's Improved Brunswick.* The best second-early. Heads large, flat, solid, and of excellent quality; stems very short. Price 5 cts.; oz., 15 cts.

Cabbage, *Early Wakefield.* The great favorite with market gardeners; the earliest, and sure to head. The seed is true and the best. Price 5 cts.; oz., 15 cts.

Cabbage, *Early Summer.* This variety is deservedly popular with market gardeners. It forms large, solid, round, flattened, compact heads, of excellent quality. The veins of the leaf are very white and prominent; it is the most compact grower of the second-earlies; a sure header. Price 5 cts.; oz., 15 cts.

Cabbage, *Stone Mason.* Very popular market variety, for fall and winter use; a sure header. Price 5 cts.; oz., 20 cts.

Cabbage, *Premium Flat Dutch.* One of the best winter sorts in cultivation; of fine quality. Price 5 cts.; oz., 15 cts.

Cabbage, *Large Red Drumhead.* Larger heads than the old "Red Dutch." Heads round in shape, of deep color, and remarkably hard and solid. Either for pickling or table use there is no better variety of red Cabbage. Price 5 cts.; oz., 15 cts.

Cabbage, *Savoy.* The best of all for home use. The plants are vigorous, very sure heading, uniformly savoyed; and heads globular. Price 5 cts.; oz., 20 cts.

Cabbage. *Autumn King.* Also known as "World Beater." It is uniformly true to type and sure to head hard—"solid as a rock." The large, broad heads are very thick through, slightly rounded at the top; fine grained and tender—more so than any other large Cabbage. It is a rapid grower, and while it will well repay good treatment, yet it does not require special culture to develop. If you want first premium at fair, for largest Cabbage, try this. Price 5 cts.; oz., 25 cts.

Cauliflower, *Early Snowball.* Probably grown more than all other varieties together. It is exceedingly early and hardy, and is one of the surest to make a solid, compact head; also, more and more in favor for planting for the late summer and fall crop. We have best seed procurable, and confidently recommend it as equal to that offered by anyone. Price, 1-2 pkt., 15 cts.; 1 pkt , 25 cts.; oz., $2.25.

Cucumber, *Early Frame.* The well known early sort. Price 5 cts.; oz., 10 cts.

Cucumber, *Nichol's Medium Green.* We consider this the most valuable sort. As a pickle sort, "Nichol's Medium Green" will be found unequalled, and for early forcing purposes or for slicing there is no better variety. It is exceedingly productive, of medium size, and always straight and smooth. The color is dark green, the flesh tender and crisp. Price 5 cts.; oz., 10 cts.

Cucumber, *Imp. White Spine.* One of the most popular either for slicing or for pickles; abundant bearer. Price 5 cts.; oz., 10 cts.

Cucumber, *Emerald.* This beautiful new Cucumber is a seedling of the "White Spine," and retains all the good points of its parent, with the addition of a rich, dark-green skin, entirely free from spines. It is strictly an evergreen, retaining its color until fully ripe. The long, straight, handsome fruits are produced early and abundantly. It is almost entirely free from "warts and spines." Most excellent, both for table and market.   Price 5 cts.; oz., 15 cts.

Cucumber, *Cool and Crisp*. This variety is the earliest and most productive of all pickling Cucumbers. At the pickling stage they are straight, long, even and slim, and until fully grown are of a dark green color. While it is one of the best pickling varieties, it is also excellent for slicing, the Cucumbers, when fully grown, being of good size and exceedingly tender and crisp; as early as "Early Frame," of better quality, and bears the whole season.   Price 5 cts.; oz., 10 cts.

Cucumber, *Long Green*. Long fruit, of excellent quality; dark green, firm and crisp. This is the improved sort, fine for all purposes. Price 5 cts.; oz., 10 cts.

Carrot, *Guerande, or Oxheart*. This variety has given great satisfaction. While not attaining one-half the length of many other varieties, it will compare favorably in bulk of crop, as on good land it will produce Carrots four to six inches in diameter. The crop can also be readily pulled, while the longer sorts require digging. One of the best for stock.   Price 5 cts.; oz., 10 cts.; lb., 60 cts.

Carrot, *Imp. Long Orange*. Too well known to need description; an immense yielder; best for stock.   Price 5 cts.; oz., 10 cts.; lb., 60 cts.

Carrot, *Chantenay* (or Model Carrot.) For table use, it is probably the best in shape and finest in quality of all. They are a little longer than the "Short Horn," being thicker at the shoulder, and hence more productive; always very smooth and fine in texture, and easily dug; the flesh is of a beautiful, deep golden orange; very tender.   Price 5 cts.; oz., 10 cts.; lb., 60 cts.

Carrot, *Danvers*. Grown largely everywhere, on account of its great productiveness and adaptability to all classes of soils. Tops medium size, coarsely divided; roots deep orange, large, but medium length, tapering uniformly to a blunt point, smooth and handsome; flesh sweet, crisp, tender, and of a deep orange color. This variety, although the roots are shorter, produces as large a bulk as the older field sorts, and is easier harvested.   Price 5 cts.; oz., 10 cts.; lb., 60 cts.

Catnip.   Price 5 cts.; oz., 25 cts.

Celery, *Golden Self-Bleaching* and *White Plume*. These two fine varieties have their inner stalks bleached at all stages of growth, so that slight banking for a few days gives a perfect bleach. The "White Plume" is earliest, but the "Golden" is one of the finest of all.   Price, either sort, 5 cts.; oz., 25 cts.

Celery, *Pink Plume*. Is practically identical with "White Plume," but with the added merit of the stalks being exquisitely suffused with pink and possessing the crispness, rich nutty flavor and long keeping qualities for which red Celeries are noted.   Price 5 cts.; oz., 25 cts.

Celery, *Giant Pascal*. The best for winter use. The stocks are remarkably large, thick, solid and entirely stringless. It is the largest Celery grown as regards width and thickness of stalks. When fully grown, the outer stalks average two inches wide, and are nearly as thick as a man's finger. It is the best of all Celeries for January and February use.   Price 5 cts.; oz., 25 cts.

Celery, *Boston Market*. Dwarf white, excellent quality. One of the best for late fall and winter use.   Price 5 cts.; oz., 25 cts.

Cress, *Water*. One of the most appetizing, delicious flavored of small salads; it does fairly well in very moist situations, but thrives best in shallow water on the edges of streams, where it grows most luxuriantly. It is a hardy perennial and increases in growth from year to year.   Price 10 cts.

Dandelion. Large leaf, new French sort, double the size of common variety. Sown in spring or summer, fit to cut following spring. Price 5 cts.; oz., 40 cts.

Egg Plant, *New York Spineless.* Fruit of largest size, and quite free from spines; of splendid quality. Price 10 cts.

Fennel. The leaves boiled, enter into many fish sauces. Price 5 cts.

Gourd, *Sugar Trough.* These gourds are useful for many household purposes, such as buckets, baskets, nest-boxes, soap and salt dishes. They grow to hold from four to ten gallons. Price 10 cts.

Gourd, *Dipper.* This variety is also useful for many household purposes. When grown on the ground the stems will be curved, while if raised as a climbing vine, on a trellis, the weight of the blossom end will cause the stem to grow straight. Price 10 cts.

Gourd, *Dishcloth or Luffa.* A natural dishcloth and a most desirable one, is furnished by the peculiar lining of this fruit, which is sponge-like, porous, tough, elastic, and durable. They are also most useful in the bath, in place of sponges. Price 10 cts.

Gourd, *Japanese Nest Egg Gourd.* These exactly resemble, in color, shape and size, the eggs of hens, do not crack, and are uninjured by cold or wet. They make the very best nest eggs. The plant is a rapidly growing climber, and is useful for covering screens, etc., being quite ornamental, with the numerous white eggs. Price 10 cts.

Gourds (mixed or ornamental). This mixture contains the following sorts: *Gooseberry, Club, Nest Egg, Lemon, Bottle, Onion, Orange, Pear, Serpent, Sugar Trough, Dipper,* and *Dishcloth.* Price, mixed, 10 cts.

Kale, *German Dwarf Curled.* Sown in spring for summer use, and in fall for spring. A very hardy variety. Price 5 cts.; oz., 10 cts.

Lavender, *Lavendula Vera.* An aromatic, medicinal herb. Price 5 cts.

Lettuce, *Iceberg.* There is no handsomer or more solid Cabbage Lettuce in cultivation—in fact, it is strikingly beautiful. The large, curly leaves which cover the outside of the solid heads are of a bright, light green, with a very slight reddish tinge at the edges. Price 5 cts.; oz., 15 cts.

Lettuce, *Deacon.* While this does not make as large a head as some others, the heads formed are very solid and of excellent flavor, and remain in condition for use as long as any variety in cultivation. Plant medium size, with very few outer leaves; leaves light green, very thick. This is certainly one of the best varieties for summer use. Price 5 cts.; oz., 15 cts.

Lettuce, *Mignonette.* A small, sturdy, russet-colored Lettuce, distinct and unique, and of great value. The entire plant measures only seven inches in diameter, and is more solid and compact than any existing variety. The outer leaves are few, and these cling so closely to the head that it is almost all head. The few outside leaves are as deeply crumpled as a Savoy Cabbage, and richly colored with russet-red and bronze-green of varying shades. The heart is of a creamy yellow, waved with pale greenish white. It is wonderfully hardy, resists drought and excessive moisture—a most valuable feature in our climate. Price 5 cts.; oz., 15 cts.

Lettuce, *Thick Head, Yellow.* Every plant heads, and the heads are uniformly shaped, very large and thick through, and of a solidity equal to that of a Cabbage. The leaves are slightly crumpled, of a beautiful greenish yellow color, and for tenderness it is unsurpassed by any existing variety. Price 5 cts.; oz., 15 cts.

Lettuce, *Perpetual.* Our customers will be especially pleased with this variety for a summer Lettuce, as it is of finest quality and remains for a long season tender and crisp; unexcelled as a long standing sort, combining tenderness, beauty and delicate flavor; best of its season. Price 5 cts.; oz., 15 cts.

Lettuce, *Prize Head.* An early sort, deep green, so washed with red as often to appear more red than green; forming a head of a dense mass of leaves rather than one like a Cabbage, and very slow to run to seed; leaves large, nearly round, frilled at the edge, and densely blistered. They are exceedingly crisp, tender and good flavored, and are not bitter as early as other varieties.   Price 5 cts.; oz., 15 cts.

Lettuce, *Black Seeded Tennis Ball.* Early hard head; few outside leaves; grown largely under glass.   Price 5 cts.; oz., 10 cts,

Lettuce, *Hanson.* Improved Hard Heading Stock. Heads green outside and white within; grows to a remarkable size, very solid, and is refreshingly sweet, crisp and tender. It withstands the hot sun. Most of the heads are so very solid that they do not send up any seed stalks unless cut open with a knife. Price 5 cts.; oz.,10 cts.

Lettuce, *Boston Curled.* Ornamental; fair quality. Price 5 cts.; oz., 10 cts.

Lettuce. Above eight varieties mixed. A fine variety for a small amount of money. You get early, medium and late in one package. Price 5 cts.; oz., 15 cts.

NOTE.—We call attention to our fine varieties of Lettuce; in growing, give good soil and plenty of room; small kinds six to eight inches, and large sorts ten to twelve inches apart.

Mustard, *White.* Used for salad and greens.   Price 5 cts.; oz., 10 cts.

Mustard, *Chinese.* Leaves twice the size of the ordinary white Mustard; stems more succulent: flavor pleasantly sweet and pungent. Price 5 cts.; oz., 10 cts.

Marjoram, *Sweet.* For seasoning.   Price 5 cts.

Musk Melon, *Emerald Gem.* All who have grown it unite in pronouncing it unequalled in rich, delicious flavor. The skin, while ribbed, is smooth and of a very deep emerald green color. The flesh, which is thicker than in any other Melon of its size, is of a suffused salmon color, and ripens thoroughly to the extremely thin green rind; luscious beyond description; altogether unapproached in delicious flavor. The Melons mature extremely early; of superb quality. Price 5 cts.; oz., 10 cts.

Musk Melon, *Melrose.* Very hardy, strong, and vigorous grower; the vines remaining green longer than any other variety. The Melons are produced abundantly, and very close to the hill—a feature of the greatest value. The color is rich, dark green; the shape is oval. The Melons average in weight about four pounds, and are finely and densely netted, showing but slight tendency to rib; the form is remarkably perfect and true. The flesh is very thick and solid; clear, light green in color, shading to a rich salmon at the seed cavity. In flavor the "Melrose" cannot be equalled by any other variety, and carries its superior quality and sweetness to the extreme edge of the skin; ripens among the earliest.   Price 5 cts.; oz., 20 cts.

Musk Melon, *Paul Rose* (or Petosky). A cross of the Osage with the Netted Gem. It possesses the fine netting of the Gem, and averages about the same size, but owing to the great thickness of its beautiful salmon-colored flesh will weigh at least one-third more than a Netted Gem of the same size. The size is much more convenient for packing in baskets than the Osage, while as a shipper and long keeper it surpasses all other red-fleshed Melons. In Chicago and some other Western cities, where there is a good demand for the red or salmon-fleshed varieties, the Paul Rose sells at double the price of other sorts. Market gardeners who grow red-fleshed Melons will find the Paul Rose their most profitable sort, while for the home table nothing better could be desired or obtained.   Price 10 cts.

Musk Melon, *Netted Gem.* The best early small green-fleshed Melon known today. Shape is almost a perfect globe. They grow remarkably uniform, weighing from one and a quarter to one and a half pounds each. They are thick-meated, the flesh is light green in color, and uniformly of fine, luscious flavor; skin green, regularly ribbed and thickly netted. Very prolific and extra early in ripening—no variety being earlier. Sure to ripen in nearly all sections.　　　Price 5 cts.; oz., 10 cts.

Water Melon, *Cole's Early.* This is one of the finest early Water Melons; of medium size, and it is so early that it matures in every state; very hardy; the flesh is deep red in color, clear to the rind, and is most sweet and delicate in flavor. It is exceedingly brittle, and hence cannot be shipped to distant markets, but is most desirable for the home garden.　　　Price 5 cts.; oz., 10 cts.; 1-4 lb., 25 cts.

Water Melon, *Mountain Sweet.* Fruit oblong, dark green; rind thin; flesh red, solid and sweet. One of the earliest large Melons.　　　Price 5 cts.; oz., 10 cts.

Water Melon, *Phinney's Early.* Flesh deep red, of very superior quality; early and productive. One of the best extra early Melons.　　　Price 5 cts.; oz., 10 cts.

Okra, *Dwarf Prolific.* A very fine new variety, much dwarfer than the ordinary dwarf Okra, and much more productive; short thick pods. Very palatable when stewed and served as is a dish of Asparagus. It is universally used throughout the South, and is as easily raised as a weed. So delicious is the flavor that when once used it will be grown every year.　　　Price 5 cts.; oz., 10 cts.

Onion, *Australian Brown* (new). Is of medium size, wonderfully hard and solid, extremely early in ripening, and never makes any stiff-necks or scallions. Every seed seems to produce a good-sized Onion, and the bulb begins to form very quickly—when the plant is not over three inches high. Planted with the Red Wethersfield, it proved to be nearly four weeks earlier, and ripened off more uniformly. From its firmness and hardness, it will keep in good condition longer than any other Onion known. The color of the skin is a clear amber-brown. Widely contrasted from any other Onion.　　　Price 10 cts.; oz., 20 cts.

Onion, *Globe Yellow Danvers.* This variety is now the standard medium yellow Onion of our markets the country over. Our strain of this sort is unsurpassed. It is early, large and a very heavy yielder. It has a very small neck and is an excellent keeper. Every grower knows the value of northern New England grown seed over all others.　　　Price 5 cts.; oz., 15 cts.; 1-4 lb., 50 cts.; 1 lb., $1.50, postpaid.

Onion, *Wethersfield Large Red,* one of the best; all our stock is New England grown, and of the very best quality. Price 5 cts.; oz., 15 cts.; lb., prepaid. $1.50.

Onion, *Early Red Globe.* Earliest of all.　　　Price 5 cts.; oz., 15 cts.

Parsley, *Double Curled.* Very nice.　　　Price 5 cts.

Parsley, *Moss Curled.* New, extra dark. When its beauty, and the tender delicacy of its flavor is known, it will undoubtedly become an extremely popular plant. It is very handsome and ornamental in growth, far more so than many plants which are grown in our flower gardens for decorative purposes. The leaves are beautifully curled or crimpled, and of a peculiar, extra dark green color. It is very productive, and from the densely-curled character of its leaves, green seasoning or garnishing can be prepared in much less time than the open-leaved varieties.　　　Price 5 cts.

Parsnip, *Guernsey.* A half long variety; fine quality. Price 5 cts.; oz., 10 cts.

Parsnip, *Hollow Crowned.* Best variety; long, smooth, white. The great market sort.　　　Price 5 cts.; oz., 10 cts.

Pepper, *Bell,* *Squash* and *Cayenne.*　　　Price, each, 5 cts.; oz., 25 cts.

Pepper Grass, *Extra Curled.* Best variety.　　　Price 5 cts.; oz., 10 cts.

Radish, *Early Long Scarlet.* Tender variety; fine flavor and crisp. One of the best for market or home use.　　　Price 5 cts.; oz., 10 cts.

Radish, *Early Round Dark Red.* Of very deep, dark red skin; beautiful shape, and extra early. This is the very best strain of Extra Early Scarlet Turnip Radish in cultivation; extra fine quality; always crisp.　　　Price 5 cts.; oz., 10 cts.

Radish, *Scarlet Turnip, White Tipped.* A small, round turnip-shaped Radish, of bright scarlet, distinctly tipped with clear white, sharply delineated, which gives a very pleasing appearance to the bunches when ready for market. This Radish ready to pull twenty days after sowing.   Price 5 cts. ; oz., 10 cts.

Radish, *Loug White Vienua.* This new Radish is the finest Long White Radish in cultivation. It is most beautiful in shape, white in color; both skin and flesh are pure snow white; crisp, brittle, and of rapid growth.   Price 5 cts. ; oz., 10 cts.

Radish, *Wood's Early Frame.* A long, red Radish, not quite so long as the "Long Scarlet," and with smaller top; it is quite ten days earlier than the "Long Scarlet" Radish in the open ground, while from its exceedingly small top it is most excellent for forcing.   Price 5 cts. ; oz., 10 cts.

Salsify, *Vegetable Oyster Plant.*   Price 5 cts. ; oz., 10 cts.

Salsify, *New Mammoth.* Grows uniformly to an extra large size, averaging fully double the size and weight of roots of the old variety. The roots, notwithstanding their enormous size, are of superior quality, and delicate flavor. Price 10 cts.

Spinach. For greens.   Price 5 cts. ; oz., 10 cts.

Sage and Summer Savory.   Price, each, 5 cts. ; oz., 20 cts.

Squash, *New Giant Summer Crookneck.* This new strain is even earlier than the common variety, while the Squashes grow to an extra large size, measuring from eighteen inches to two feet long, and of finest quality.   Price 5 cts. ; oz., 15 cts.

Squash, *Summer Crookueck.* Best summer sort.   Price 5 cts. ; oz., 10 cts.

Squash, *Hubbard.* Well known winter sorts. More of this variety grown than of all other winter squashes combined.   Price 5 cts. ; oz., 10 cts. ; lb., 65 cts.

Squash, *Essex Hybrid.* Early, yet a long keeper; flesh dry, thick and sweet, better than "Turban;" a standard variety.   Price 5 cts. ; oz., 10 cts. ; lb., 65 cts.

Squash, *Faxon* (or the Squash of many colors). The fixed characteristics are earliness, long keeping, uniformity of shape, variety of colors, and the excelling in richness, sweetness and flavor. It is very productive, of medium size, and is destined to become a standard among Squashes, both for home use and for general marketing. Even when not fully ripe it is of excellent quality as a summer Squash. It is a good keeper, and has the exceptional recommendation of being a table delicacy through three seasons—summer, fall and winter. For sweetness and dryness, the "Faxon" Squash cannot be excelled, if equalled,   Price 5 cts. ; oz., 10 cts. ; lb., 75 cts.

Squash, *Dunlap's Early Prolific Marrow.* The standard early of the running varieties. No variety can compete with it for earliness, it being about twelve days ahead of the "Boston Marrow," and for this reason can be planted a week later than any other kind. It grows to weigh from seven to twenty pounds, and is very productive. Its color is most attractive, a brilliant orange-red. Quality excellent when mature. A good keeper.   Price 5 cts. ; oz., 10 cts. : lb., 65 cts.

Squash, *Bush Scallop.* Best early of the scalloped sorts. Quality not considered equal to the "Summer Crookneck."   Price 5 cts. ; oz., 10 cts.

Squash, *Warted Hubbard.* Same as "Hubbard," except warty knobs all over it. Quality the best.   Price 5 cts. ; oz., 10 cts.

Tomato, *Early Bird* (new). Extremely early, and said to be superior to any of the first early class and a great yielder. Its great earliness, fine size, smoothness, firmness, beautiful bright skin and wonderful productiveness are far beyond what the most enthusiastic growers ever expected to realize.   Price 10 cts.

Tomato, *Fancy.* A compact, bushy form ; foliage dark green, heavily blistered or crimped, as in a fine Savoy Cabbage. The fruits are of medium size, very smooth and regular both in size and form, solid and free from cracks, quite fleshy, with few seeds, and of sweet flavor. When ripened under the warm sun they are a rich purplish-red. It is characteristic of the variety to produce fruit rather than vines. It does not send out laterals readily, but seems to put all efforts toward perfecting and ripening the fruit, and when the laterals are sent out they grow the fruit set on them rather than seek on growing to vines. Price 10 cts.

Tomato, *The Aristocrat.* An excellent new sort, resembling in habit of growth and foliage the " Dwarf Champion," but the color of the fruit is a fine rich glossy red ; solid and smooth, and in size from medium to large. Price 5 cts. ; oz., 25 cts.

Tomato, *Dwarf Champion.* The " Dwarf Champion " Tomato is one of the most valuable sorts introduced in many years. It is entirely distinct in habit of growth and foliage from any other sort. It is dwarf and compact in habit, the plant growing stiff and upright, with very thick and short-jointed stems. In a comparative test with leading varieties it has proven itself remarkably early. Always smooth, symmetrical and attractive in appearance; flesh solid and ripens well. Price 5 cts. ; oz., 25 cts.

Tomato, *Matchless* (Improved Trophy). The fruits are of the largest size, and the size of the fruit is maintained throughout the season, the healthy growth of foliage continuing until killed by frost. Had we to confine ourselves to one variety, it would be " The Matchless," for certainly no other Tomato will produce extra large fruits, so smooth, handsome and marketable. The skin is remarkably tough and solid, so that ripe specimens picked from the vine will keep in good marketable condition for two weeks. Its fine quality, solidity, and long-keeping character have already made it a great favorite for market and family use. So great an improvement on the "Trophy " that we drop that variety. Price 5 cts. ; oz., 25 cts.

Tomato, *Yellow Plum.* Small yellow variety; very early. Price 5 cts.

Tomato, *Perfection.* Fine smooth variety. Price 5 cts. ; oz., 25 cts.

Turnip, *Golden Ball* (or Orange Jelly). This is the most distinct yellow Turnip we know. The flesh is of very fine texture, making it one of the best table varieties. Beautiful color; fine flesh. The bulb is of medium size, with small tap roots, and is early in maturing. Price 5 cts.; oz., 10 cts.; lb., 60 cts.

Turnip, *Yellow Stone.* One of the best and most popular yellow-fleshed varieties for table use; equally good for feeding stock. Price 5 cts. ; oz., 10 cts. ; lb., 60 cts.

Turnip, *Strap Leaf.* " Red Top " or " White Top," well known early sorts.
Price, either sort, 5 cts.; oz., 10 cts.; lb., 55 cts.

Turnip, *Purple Top White Globe.* An early variety, globe-shaped; heavy cropper, in other respects similar to the " Red Top Strap Leaf." A handsome-looking bulb, and is rapidly taking the lead of all other varieties of early Turnips for market garden purposes. Price 5 cts.; oz., 10 cts.; lb., 60 cts.

Turnip, *White Egg.* Its shape is nearly oval or egg; flesh very firm and fine grained, thin and perfectly smooth skin, both flesh and skin are of snowy whiteness. Its flavor is of the best. Price 5 cts.; oz., 10 cts.; lb., 60 cts.

Turnip, *Sweet German.* The well known winter variety. Very sweet; extra for table use. The leading sort for winter. Price 5 cts.; oz., 10 cts.; lb., 55 cts.

Turnip, *Laing's Swede.* Quick growing; good for table or for stock; yellowish flesh. Price 5 cts. ; oz., 10 cts. ; lb., 60 cts.

Turnip, *Scarlet Kashmyr.* As early as the earliest known sort; most distinct in appearance, small, rarely exceeding three inches in diameter. The outer color of the root is unique, being a rich scarlet, verging on crimson ; the interior color is pure sparkling white. Price 5 cts.; oz., 15 cts.

Turnip, *Burpee's Bread-Stone.* This new Turnip is of medium size, very quick growing and the roots are very smooth and white with a faint green top. The flesh is perfectly white, fine grained, remarkably tender, very sweet; it is the finest Turnip for winter and spring use we have ever seen. It is good when harvested in the fall, but its comparative merits do not fully show up until spring, when other varieties grow pithy, strong and uneatable.   Price 5 cts.; oz., 10 cts.

PEAS.

Prepaid at prices given. Packages are of good size, containing two and one-half ounces.

American Wonder. The earliest of all wrinkled Peas. Of dwarf habit, growing from nine to twelve inches high, and producing a profusion of good sized and well filled pods of the finest flavor.
Price, pkt., 10 cts.; pint, 25 cts.; quart, 40 cts.

Premium Gem. A valuable variety, growing about fifteen inches high, and among the earliest dwarf, green, wrinkled sorts. The quality is unsurpassed; the most remarkable characteristic is its wonderful prolificness. Superior to "Little Gem." Price 10 cts.; pint, 25 cts.; quart, 45 cts.

Nott's Excelsior. One of the best of recent introduction; an improvement on "AmericanWonder," being more vigorous and prolific, with larger pods and more Peas. A decided acquisition, and sure to become popular when generally known; height, one foot. Price, pkt., 10 cts.; pint, 25 cts.; quart, 40cts.

Stratagem (improved). The pods are of immense size and uniformly filled with very large dark green peas of the finest quality. Vine medium height, but very stocky, with very broad, light green leaves, and bearing an abundance of large handsome pods. One of the very best varieties for market gardeners or home use; medium height. Price, pkt., 10 cts.; pint, 25 cts.; quart, 45 cts.

Yorkshire Hero. Long, round pods, closely filled with large, luscious, wrinkled peas; very productive. Well worthy of general cultivation; of similar type to the well known "Champion of England;" in many sections it is better and more productive. Price, pkt., 10 cts.; pint, 20 cts.; quart, 40 cts.

Champion of England. Undoubtedly one of the best Peas grown, and very popular. It is very productive, peas of a delicious flavor; height, four to five feet. Price 10 cts.; pint, 20 cts.; quart, 35 cts.

Large Black Eye Marrowfat. An excellent variety, growing about five feet high; pods large; a prolific bearer; one of the very best "Marrowfat" sorts. Price, pkt., 10 cts.; pint, 20 cts.; quart, 35 cts.

Sunol. Mr. J. H. Allen, who has originated more varieties of new Peas than any seed grower on this side of the Atlantic ocean, sends this newcomer out as his latest improvement in the early market garden sorts. He claims that it is the earliest of all to mature, and is decidedly the most profitable to raise of all the early varieties. J. J. H. Gregory says of it—"I find 'Sunol' leads them all in earliness; medium height." Price 10 cts.; pint, 20 cts.; quart, 40 cts.

First of All. A very early market sort; height about two and one-half feet. "Sunol" is an improvement on this variety. Price 10 cts.; pint, 20 cts.; quart, 40 cts.

Queen. An excellent new wrinkled variety of English origin, growing about two and a half feet high and bearing an abundance of very large handsome pods, which are well filled with large, oval, dark green peas of the finest quality and flavor. It is recommended as one of the very best main crop sorts. Price 10 cts.; pint, 25 cts.

If by express, not prepaid, 15 cts. per quart may be deducted from the above prices. Special prices given to buyers in quantities of one-half peck or more of a kind.

BEANS.

Extra Early Golden Cluster Wax (Pole). Long golden-yellow pods, borne in clusters; extra early, stringless, snaps easily, vigorous, prolific, popular; quality tender, delicious. It is unequalled for earliness, productiveness and fine quality. The pods are almost round, entirely stringless, remarkably rich, tender and buttery when cooked; of the best possible flavor. The pods are fit to use when only two or three inches long; keep their fine, tender, rich and buttery qualities until fully ripe. The most productive Bean in cultivation. Price, postpaid, 10 cts.; 1-2 pint, 15 cts.; pint, 25 cts.

Horticultural Pole or Pole Cranberry. Standard sort. Price, pkt., 10 cts.; pint, 25 cts.

Yosemite. Mammoth Wax Bush Bean. The pods frequently attain a length of ten to fourteen inches, with the thickness of a man's finger, and are nearly all solid pulp. The pods are a rich golden color, and are absolutely stringless, cooking tender and delicious. It is enormously productive, as many as fifty of its monster pods having been counted on one bush. Price, pkt., 10 cts.; pint, 25 cts.

Golden Wax (Bush). A strong growing, distinct variety, at least a week earlier than the "Black Wax." The pods are long, brittle, and entirely stringless. As a snap Bean it perhaps excels all others in tenderness and richness of flavor. The finest of all as a string Bean. Price, pkt., 10 cts.; pint, 25 cts.

Dwarf Horticultural or Bush Cranberry. One of the finest bush varieties for a green shell Bean. Price, pkt., 10 cts.; pint, 20 cts.; quart, 40 cts.

Davis's White Kidney Wax. A new variety. Plant vigorous, and of compact upright growth, with pods extra long, straight, oval, of a clear, waxy, white color, often seven to eight inches in length, and when fit for use, quite stringless. Price, pkt., 10 cts.; pint, 20 cts.

White Case-Knife. Early pole bean,with pods very long, flat, irregular, green, changing to cream white, useful both as a shell and a string Bean. Price, pkt., 10 cts.; pint, 20 cts.

Rust-Proof Golden Wax. This improved strain is destined to supercede the well-known "Golden Wax," which it has greatly out-yielded in comparative trials. The straight, handsome pods are thicker through, superior in quality; claimed by originator to be rust-proof. Price, pkt., 10 cts.; pint, 25 cts.

Henderson's Bush Lima. Fit to use from two to three weeks earlier than any other variety of the climbing or bush Limas. It grows about eighteen inches high and produces immense crops of delicious Lima Beans, continuing to bear until cut down by frost. Price, pkt., 10 cts., pint, 20 cts.

Burpee's Bush Lima. The only bush form of the true delicious large Lima. It is pronounced by all good judges as unquestionably the real bush Lima, distinct and superior in every way—in its splendidly vigorous, upright, bushy habit of growth, great uniformity and trueness to type, and always bearing the delicious large Lima beans in great abundance. This requires the same season for growth as the large pole Lima, and can be grown only where that variety succeeds. Price, pkt., 10 cts.; pint, 25 cts.

Long Yellow Six Weeks. Early, very productive and of excellent quality. Pods are often eighteen inches long. Seeds, when ripe, yellow or dun color. Price, pkt., 10 cts.; pint, 20 cts.

King of Garden. Large pole Lima; a grand Bean, but rather late for northern gardens. Price 10 cts.; pint, 25 cts.

SWEET CORN.

First-Crop. Superior to all other very early sorts, larger, earlier and very much sweeter; of dwarf habit; kernels white; ears eight-rowed and of good size. The sweetest sort for first crop. This sort is similar to "First of All," except in quality, which is so much superior in sweetness to that sort that we drop it from list. First-Crop is the best very early Corn we have ever grown.

Country Gentleman (new). The "Country Gentleman" is the finest of all sweet Corns, and will delight the most fastidious epicure, retaining its delicate tenderness and flavor even when a little old. For private family use, where quality is of the first consideration, it has no equal. The ears are of good size, and are produced in great abundance, frequently bearing four good ears, while the average is three ears to a stalk. The cob is very small, giving great depth to the kernels, which are of pearly whiteness. But the great merit of the "Country Gentleman" Corn is its delicious quality.

Cory. A standard extra early sort; good for home use and extra for market.

Early Shaker. Fine to follow "Cory," very sweet; good for home or market.

Black Mexican. Medium early, good to follow "Shaker;" one of the sweetest.

Evergreen. The best late sort; large, sweet and fine.

Squantum. One of the sweetest varieties, and is largely used for market and canning. It is a general favorite, and is wonderfully productive. The "Squantum" is the variety used almost exclusively at the famous Rhode Island clambakes, which is sufficient evidence of its quality. This and "Potter's Excelsior" are identical.

Crosby's Early. Early and a great favorite. Rather small ears, but productive and of excellent quality; a favorite for family use.

Price, corn, large packets, 10 cts.; pint, 25 cts.; quart, 45 cts.; prepaid. By express, (not prepaid), pint, 15 cts.; quart, 25 cts.

POP CORN.

Queen's Golden. The stocks grow six feet high, and the large ears are produced in abundance. Its quality and handsome appearance when popped are very noticable. It pops perfectly white, and a single kernel will expand to a diameter of nearly one inch. Price, 10 cts. per pkt.; 25 cts. per pint.

White Rice. The most widely known variety; very popular for parching. Selected seed. Price, per pkt., 10 cts.; pint, 25 cts.

VEGETABLE PLANTS AND ROOTS.

Holt's Mammoth Sage. Plant of strong growth on rich ground, often attaining a diameter of three feet the first season; leaves which are of immense size, are borne on strong stems; in quality it is of unusual substance and of strong flavor. Give it rich, good culture, and you will be astonished at the large quantity of superior Sage which can be cut from a single plant. It rarely flowers, and has never seeded in our climate. No seed to offer. Price, plants, 10 cts.; three for 25 cts.

Tomato Plants. *Dwarf Champion, Perfection, Aristocrat, Matchless, Yellow Plum, Early Bird and Fancy.* Price, pot plants, 6 cts.; 60 cts. per doz.; hot-bed plants, 45 cts. per doz.

Pepper Plants. Price 3 cts.; 30 cts. per doz.

Cauliflower Plants, *Early Snowball.* Price, doz., 20 cts.; 100, $1.50.

Celery, *White Plume, Golden Self-Bleaching, Boston Market, Giant Pascal.* Transplanted plants. Price, doz., 15 cts.; 100, 75 cts.

Asparagus Roots, *Conover's Colossal.* Fine 2-year roots. Price, 100, $1.00, by express only.

Cabbage Plants. Early plants before June 10. Price, doz., 15 cts.; 100, $1.00.

Cabbage Plants. *Fottler's, Stone Mason, Winningstadt, Premium Flat Dutch, All-Head, Savoy.* Field grown after June 15. Price, doz., 10 cts.; 100 40 cts.; 1,000, $2.00.

Mint, *(Mentha Veridis.)* This is the Mint that is so much used for culinary purposes; also known in many places as "Spearmint." Immense quantities now grown under glass to supply the large city markets. Price, two for 15 cts.; five for 25 cts.; doz., 50 cts.

Tarragon. Now appreciated by all who know it for the use of its aromatic leaves in seasoning or salads, also for Tarragon vinegar. The foliage, if cut in autumn, can be kept in a dry state the same as other herbs. In the Northern States give some protection during the winter. A teaspoonful of minced Tarragon may properly be added to any salad, dressed with oil and vinegar. Price 15 cts.

SMALL FRUITS.
STRAWBERRIES.

Michigan (new). Perfect flowering, and undoubtedly the best late Strawberry ever introduced, and for profitable market growing is unexcelled by any other, coming into market as most other kinds are going out. The berries are large, very uniform, of deep crimson color, firm and handsome. The plant is a strong healthy grower, with luxuriant foliage, and exceedingly productive.

Clyde. Fruit very large, light red, conical, firm and best quality. Plant very vigorous, plenty of runners, hardy, free from rust and very productive. It may be called a perfect variety. It resists drouth on account of its habit of deep rooting; grows equally well on any soil. A perfect flower; color, light scarlet; quality the best, and the most productive Strawberry known. Season, medium to quite late. It is called the business berry.

Lovett's Early. Of superior quality, excellent form, very uniform in both size and shape throughout the season, bright crimson color, has a perfect blossom and the most enduring healthy foliage. It holds its size to the close of the season better than other varieties, maintaining perfect vigor until all berries have ripened; berries color all over at once.

Price, either variety, prepaid, 25 cts. per doz.; by express not prepaid, $1.00 for 100.

CURRANTS.

Fay's Prolific. The leading red variety, and it is one of the best Currants we have. It has been widely planted, and has given general satisfaction. The bush is a strong grower, wonderfully prolific. Price, 1-year, 10 cts., $1.00 per doz.; 2-year, 15 cts., $1.50 per doz. Strong, healthy bushes.

White Grape. This is a Currant that should be in the family garden, not only for its handsome appearance, but for its fine quality. Its bunches are extremely long, berries large, of a beautiful translucent white, and excellent flavor; the best white variety. Price, 1-year, $1.00 per doz.; 2-year, $1.50 per doz.

RASPBERRIES.

Loudon. This is indeed a very valuable variety. A strong grower with beautiful foliage, exceedingly prolific and perfectly hardy. The berries are very large and beautiful; the richest and finest in quality of any entirely hardy and reliable variety. Price 3 for 25 cts.; 75 cts. per doz.

Cuthbert. The leading late market variety. The canes are hardy and of strong, rampant growth, with large, healthy foliage, and exceedingly productive. Berries large, crimson, and of good flavor. More largely grown than all other red Raspberries combined. Price 75 cts. per doz.

Lovett (black). Of ironclad hardiness, and is the strongest in growth of cane of any. In enormous yield it is without an equal. Add to these properties superior quality, jet black color, firmness and long life after gathered, adhering to the bush when ripe, and above all, its earliness, and we have in it what has so long been wanted, and a most valuable fruit. Distinct from all other sorts. Price 75 cts. per doz.

"Tyrian" Plant Sprinkler. Readily appreciated. It is simple in construction, consisting only of a rubber bulb and a hard rubber spray, so cannot easily get out of order. You can sprinkle your plants or flowers very quickly without injuring them or soaking the earth, and without soiling everything else near. In fact it sprinkles in nature's own way, just like rain. May also be used for sprinkling clothes, spraying carpets and clothing to prevent moths, spraying disinfectants in the sick-room, and deodorizing. Preferable in every way to the dipper or tin watering pot. Price, 6-oz., bent neck, 75 cts., postpaid.

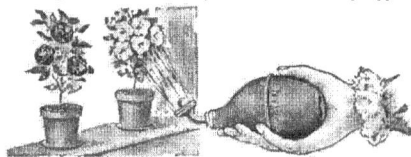

Nikoteen. It is composed of that element in tobacco which gives it its value as an insecticide and nothing else. While quite expensive, it is so concentrated that only a very small quantity is needed in spraying—one teaspoonful to one quart of water being sufficient to kill green-fly on house plants, and nearly all insects on rose bushes and shrubs in open air. (Circular upon application.) Price, per pint, $1.50.

SUMMER FLOWERINC BULBS.

Gloxinia Crassifolia. A charming class of summer-blooming bulbs, which succeed with ordinary care. They should be grown in a moderately shady place, as the sun burns the foliage when wet, making brown spots appear. They are easily grown, and bloom constantly until autumn, when they should be allowed to die down and the pots kept dry in some warm place until spring, when the bulbs can be started into growth again. Any good soil will grow them, but will do best in compost of one-half good loam, balance equal parts old manure and leaf mould. In potting, allow top of bulb to stand just above the soil—water but little until growth commences. Our bulbs this season are fine and large, and in a mixture of the grandest colors; all in the "Crassifolia" varieties, which is the strongest growing strain known. (See cut.) Price, in colors, red, white, blue, and white with blue border, large bulbs, 30 cts., the four for $1.00; fine mixed strong bulbs, 15 cts., four for 50 cts.; mixed, in mammoth bulbs, 25 cts.; three all different, but unnamed, 60 cts.

Lily, Burbank. A new wonder from California, said by the introducer to be the Lily for the millions. Very hardy; increases rapidly, soon forming a large clump, single stalks from this clump often having, if well established, over one hundred blooms. Color, yellow, with brown spots. Price 30 cts.

Lilium Auratum. The glorious gold-banded lily of Japan, and one of the grandest plants in cultivation. Its immense ivory-white flowers are thickly studded with yellow and crimson spots, while in the center of each petal is a golden band, fading at its edges into the white. Price 20 cts.

Lilium Speciosum Rubrum. No words can overstate the brilliant beauty of these famous Japan lilies. The six broad white or pink petals are thickly dotted with rose or carmine spots, and the graceful form, brilliant color and exquisite fragrance make them very effective. Price 20 cts.

Lilium Speciosum Album. Pure white flowers with a greenish band through the center of each petal. They are of great substance and very fragrant. Price 25 cts.

The above lilies are all hardy for open ground or are fine for pot culture, but should be planted not later than May 20th, if to bloom nicely this year. Our bulbs are extra large and fine.

Tuberous Begonias. Single mixed. See discription, Page 15. Price 15 cts.; 5 for 50 cts.

Tuberous Begonias. Double mixed. Price 20 cts.; 4 for 50 cts.

Caladium Esculentum (Elephant's Ear). Very ornamental. Mammoth leaves. Use plenty of water. Fine strong bulbs. Price 25 cts. prepaid.

Hyacinthus Candicans. A magnificent, Yucca-like plant, producing in July and August a flower stem three to four feet high covered with from twenty to thirty pure white, pendant, bell-shaped flowers. Price, fine bulbs, 10 cts.; 4 for 25 cts.

Gladiolus. See page 37.

Tigrida's Tiger (or shell flower). Three varieties: yellow, spotted crimson; red, spotted yellow; white, spotted crimson. Price, 3 for 15 cts.

Spotted Leaf Cullu. Dark green leaves, beautifully spotted with white; the flowers are white with a purple throat. A fine ornamental summer flowering plant. Price 15 cts.

EVERGREENS AND SHRUBS.

Norway Spruce and Arbor Vitae. Fine trees, home grown. They have been transplanted two or three times, and are in fine condition. Reduced rates in quantity or for hedging.
Price, 3½ to 4½ feet, 50 cts.; 4½ to 5½ feet, 75 cts.; 5½ to 7 feet high, $1.00 each.

Japanese Lilac. A wonder from Japan; does not sucker from root, but makes a neat small tree, blooms a month later than all other sorts; color, creamy white. Price, 2 to 3 feet, 50 cts.; 6 to 7 feet, $1.50.

Lilac, Mad. Lemoine. Double white florets like miniature tuberoses. The most beautiful lilac we have ever seen. Price, fine plants, $1.00.

Lilac, Marie Le Graye. Immense trusses of single white flowers; none equal this in size of truss; also grown as a plant for winter bloom. Price, strong bushes, 75 cts.

Duetzia, Pride of Rochester. Finest double white. Price 50 cts.

Pyrus Japonicus. Or Japan quince; red flowers. Price 50 cts.

Snow Ball. Strong bushes. Price 50 cts.

Wigelia, Variegated. Foliage green and creamy white, flowers pink. Price 50 cts.

Wigelia, Candida. Pure white flowers in greatest profusion. Price 50 cts.

Floral Department

❧ ❧

WE would call the attention of our patrons to our present facilities for furnishing seasonable choice flowers at all times. We also furnish and execute in the most modern and artistic manner, all kinds of designs, Table, House or Church Decorations, for any desired purpose. With our large and varied stock of flowers, competent artists and long experience, we can confidently place our work in competition with the best in the land. All orders intrusted to us will be executed promptly, in latest styles, and at reasonable prices. We solicit trial orders from parties living not over *eighteen* hours' distant by rail.

Address,

ELLIS BROS.,

KEENE, N. H.